Statistics for Absolute Beginners

(First Edition)

O Theobald

Published by Scatterplot Press

Please contact the author at **oliver.theobald@scatterplotpress.com** for feedback, media contact, a university desk copy, omissions or errors regarding this book.

Find Us On:

Skillshare

www.skillshare.com/user/machinelearning_beginners

Teachable

http://scatterplotpress.teachable.com/

YouTube

Scatterplot Media

Instagram

machinelearning_beginners

TABLE OF CONTENTS

General Terms

Data

A term for any value that describes the characteristics and attributes of an item that can be moved, processed, and analyzed. The item could be a transaction, a person, an event, a result, a change in the weather, and infinite other possibilities. Data can contain various sorts of information, and through statistical analysis, these recorded values can be better understood and used to support or debunk a research hypothesis.

Population

The parent group from which the experiment's data is collected, e.g., all registered users of an online shopping platform *or* all investors of cryptocurrency.

Sample

A subset of a population collected for the purpose of an experiment, e.g., 10% of all registered users of an online shopping platform *or* 5% of all investors of cryptocurrency. A sample is often used in statistical experiments for practical reasons, as it might be impossible or prohibitively expensive to directly analyze the full population.

Variable

A characteristic of an item from the population that varies in quantity or quality from another item, e.g., the **Category** of a product sold on Amazon. A variable that varies in regards to quantity and takes on numeric values is known as a quantitative variable, e.g., the **Price** of a product. A variable that varies in quality/class is called a qualitative variable, e.g., the **Product Name** of an item sold on Amazon. This process is often referred to as *classification*, as it involves assigning a class to a variable.

Product Name	Category	Price ($)	2-Day Delivery	Reviews	Size
iPhone	Phones	699	Yes	9902	Small
Air Pods	Phones	139	Yes	2340	Small
Standing Desk	Office Equipment	399	No	23	Large
Statistics for Beginners Vol. 1	Books	9.90	No	44	Small
Fitbit	Fitness	149	Yes	3300	Small

Amazon product dataset

Discrete Variable

A variable that can only accept a finite number of values, e.g., customers purchasing a product on Amazon.com can rate the product as 1, 2, 3, 4, or 5 stars. In other words, the product has five distinct rating possibilities, and the reviewer cannot submit their own rating value of 2.5 or 0.0009. Helpful tip: qualitative variables are discrete, e.g., **Name** or **Category** of a product.

Product Name	Category	Price ($)	2-Day Delivery	Reviews	Size
iPhone	Phones	699	Yes	9902	Small
Air Pods	Phones	139	Yes	2340	Small
Standing Desk	Office Equipment	399	No	23	Large
Statistics for Beginners Vol. 1	Books	9.90	No	44	Small
Fitbit	Fitness	149	Yes	3300	Small

Continuous Variable

A variable that can assume an infinite number of values, e.g., depending on supply and demand, gold can be converted into unlimited possible values expressed in U.S. dollars. Opposite to a discrete variable, a continuous variable can also assume values arbitrarily close together. In the case of our dataset, **Price** and **Reviews** are continuous variables.

Product Name	Category	Price ($)	2-Day Delivery	Reviews	Size
iPhone	Phones	699	Yes	9902	Small
Air Pods	Phones	139	Yes	2340	Small
Standing Desk	Office Equipment	399	No	23	Large
Statistics for Beginners Vol. 1	Books	9.90	No	44	Small
Fitbit	Fitness	149	Yes	3300	Small

Categorical Variables

A variable whose possible values consist of a discrete set of categories, such as gender or political allegiance, rather than numbers quantifying values on a continuous scale.

Ordinal Variables

As a subcategory of categorical variables, ordinal variables categorize values in a logical and meaningful sequence. Unlike standard categorical variables, i.e. gender or film genre, ordinal variables contain an intrinsic ordering or sequence such as {**small**; **medium**; **large**} or {**dissatisfied**; **neutral**; **satisfied**; **very satisfied**}.

Product Name	Category	Price ($)	2-Day Delivery	Reviews	Size
iPhone	Phones	699	Yes	9902	Small
Air Pods	Phones	139	Yes	2340	Small
Standing Desk	Office Equipment	399	No	23	Large
Statistics for Beginners Vol. 1	Books	9.90	No	44	Small
Fitbit	Fitness	149	Yes	3300	Small

The distance of separation between ordinal variables does not need to be consistent or quantified. For example, the measurable gap in performance between a gold and silver medalist in athletics need not mirror the difference in performance between a silver and bronze medalist.

Independent and Dependent Variables

An independent variable (expressed as X) is the variable that supposedly impacts the dependent variable (expressed as y). For example, the supply of oil (independent variable) impacts the cost of fuel (dependent variable). As the dependent variable is "dependent" on the independent variable, it is generally the independent variable that is tested in experiments. As the value of the independent variable changes, the effect on the dependent variable is observed and recorded.

	Independent	Dependent	Independent		Independent	
Product Name	**Category**	**Price ($)**	**2-Day Delivery**		**Reviews**	**Size**
iPhone	Phones	699	Yes		9902	Small
Air Pods	Phones	139	Yes		2340	Small
Standing Desk	Office Equipment	399	No		23	Large
Statistics for Beginners Vol. 1	Books	9.90	No		44	Small
Fitbit	Fitness	149	Yes		3300	Small

In analyzing Amazon products, we could examine **Category**, **Reviews**, and **2-Day Delivery** as the independent variables and observe how changes in those variables affect the dependent variable of **Price**. Equally, we could select the **Reviews** variable as the dependent variable and examine **Price**, **2-Day Delivery**, and **Category** as the independent variables and observe how these variables influence the number of customer reviews.

	Independent	Independent	Independent		Dependent	
Product Name	**Category**	**Price ($)**	**2-Day Delivery**		**Reviews**	**Size**
iPhone	Phones	699	Yes		9902	Small
Air Pods	Phones	139	Yes		2340	Small
Standing Desk	Office Equipment	399	No		23	Large
Statistics for Beginners Vol. 1	Books	9.90	No		44	Small
Fitbit	Fitness	149	Yes		3300	Small

The labels of "independent" and "dependent" are hence determined by experiment design rather than inherent composition, which means one variable could be a dependent variable in one study and an independent variable in another.

1

INTRODUCTION

"Let's listen to the data."
"Do you have the numbers to back that up?"

We live in an age and society where we *trust* technology and quantifiable information more than we trust each other—and sometimes ourselves. The gut feeling and conviction of Steve Jobs to know "what consumers would later want" is revered and romanticized. Yet there's sparse literature (*Blink* by Malcolm Gladwell is a notable exception), an eerie absence of online learning courses, and little sign of a mainstream movement promoting one person's unaided intuition as a prerequisite to success in business. Everyone is too preoccupied with thinking about quantitative evidence, including the personal data generated by Apple's expanding line of products. Extensive customer profiling and procuring data designed to wrench out our every hidden desire are dominant and pervasive trends in business today.

Perhaps Jobs represents a statistical anomaly. His legacy cannot be wiped from the dataset, but few in the business world would set out to emulate him without data in their pocket. As *Wired Magazine's* Editor-in-chief Chris Anderson puts it, we don't need theories but rather data to look at and analyze in the current age of big data.[1]

Data—both big and small—is collected instantly and constantly: how far we travel each day, who we interact with online and where we spend our money. Every bit of data has a story to tell. But, left isolated, these parcels of information rest dormant and underutilized—equivalent to Lego blocks cordoned into bags of separate pieces.

Data, though, is extraordinarily versatile in the hands of the right operator. Like Lego laid out across the floor, it can be arranged and merged to serve in a variety of ways and rearranged to derive value beyond its primary purpose. A demonstration of data's secondary value came in 2002 when Amazon signed a deal with AOL granting it access to user data from AOL's e-commerce platform. While AOL viewed their data in terms of its primary value (recorded sales data), Amazon saw a secondary value that would improve its ability to push personalized product recommendations to users. By gaining access to data that documented what AOL users were browsing and purchasing, Amazon was able to improve the performance of its own product recommendations, explains Amazon's former Chief Scientist Andreas Weigend.[2]

Various fields of data analytics including machine learning, data mining, and deep learning continue to improve our ability to unlock patterns hidden in data for direct or secondary analysis as typified by Amazon. But behind each new advanced technique is a trusted and lasting method of attaining insight, popularized more than two and a half hundred years ago under the title of this book.

While primary methods of statistical analysis date back to at least the 5th Century BC, it wasn't until the 18th Century AD that these and newly evolved methods coalesced into a distinctive sub-field of mathematics and probability known today as *statistics*.

---

[1] Viktor Mayer-Schonberger & Kenneth Cukier, "Big Data: A Revolution That Will Transform How We Live, Work and Think," *Hodder & Stoughton*, 2013.
[2] Viktor Mayer-Schonberger & Kenneth Cukier, "Big Data: A Revolution That Will Transform How We Live, Work and Think," *Hodder & Stoughton*, 2013.

A notable frontrunner to the developments of the 18th Century was John Graunt's publication *Natural and Political Observations Made upon the Bills of Mortality.* The London-born haberdasher[3] and his friend, William Petty, are credited with developing the early techniques of census analysis that later provided the framework for modern demographic studies.

Graunt developed the first "life table," which surmised the probability of survival amongst age groups during a public health crisis that hit Europe in the mid-1600s. By analyzing the weekly bills of mortality (deaths), Graunt and Petty attempted to create a warning system to offset the spread of the bubonic plague in London. While the system was never actually implemented, it served as a useful statistical exercise in estimating London's sizeable population.

Probability theory evolved during this same period courtesy of new theories published by Gerolamo Cardano, Blaise Pascal, and Pierre de Fermat. As an accomplished chess player and gambler in Italy, Cardano observed dice games to comprehend and distill the basic concepts of probability. This included the ability to produce a desired result by defining odds as the ratio of favorable to unfavorable outcomes. He subsequently wrote *Liber de ludo aleae (Book on Games of Chance)* in 1564, but the book wasn't published until a century later in 1663. Beyond its section on effective cheating methods, Cardano's thesis was well received as the first systematic treatment of probability.

A decade earlier, in 1654, Pierre de Fermat and Blaise Pascal (also known for his work on the arithmetical triangle and co-inventor of the mechanical calculator) collaborated to develop the concept of *expected value,* which was again developed for the purpose of interpreting gambling scenarios. To end a game early, Pascal and de Fermat devised a method to divide the

---

[3] A dealer in small items used in sewing, such as buttons, zips, and thread (UK definition).

stakes equitably based on the calculated probability that each player had of winning. The French duo's study into the mathematical theory of probability helped develop the concept of *expected value* or *the law of large numbers*.

Pascal and de Fermat found that as the number of independent trials increases, the average of the outcomes creeps toward an expected value, which is calculated as the sum of all possible values multiplied by the probability of its occurrence. If you continually roll a six-sided dice, for example, the expected average value of all the results is close to 3.5.

Example

$(1 + 2 + 3 + 4 + 5 + 6) \times (1/6)$

$21 \times 0.16666666666667$

$= 3.5$

By the 18th Century, further breakthroughs in probability theory and the study of demography (based on Graunt's prior work in census studies) combined to spawn the modern field of statistics. Derived from the Latin stem "sta," meaning "to stand, set down, make or be firm," the field of statistics was initially limited to policy discussions and the condition of the state.[4] The earliest known recording of the term is linked to the German word "statistik," which was popularized and supposedly coined by the German political scientist Gottfried Aschenwall (1719-1772) in his 1748 publication *Vorbereitung zur Staatswissenschaft*.[5]

The German word "statistik" is thought to have borrowed from the Modern Latin term "statisticum collegium" (lecture course on state affairs), the Italian word "statista" (statesman or one skilled in statecraft), and the Latin word "status" (meaning a station,

---

[4] "*sta," *Etymonline*, accessed July 2, 2017,
www.etymonline.com/word/statistics/.
[5] "Statistics," *Etymonline*, accessed July 2, 2017,
www.etymonline.com/word/statistics/.

position, place, order, arrangement, or condition).[6] The new term was later published in the English-speaking world by Sir John Sinclair in his 1791 publication the *Statistical Account of Scotland*.

By the close of the 18th Century, "statistics" was synonymous with the systematic collection and analysis of a state's demographic and economic information. The regular recording of state resources had been in practice since ancient times, but deeper than an exercise in state bookkeeping, the new moniker inspired specialist studies in utilizing data to inform decision-making and incorporated the latest methods in distribution and probability.

Statistics subsequently expanded in scope during the 19th Century. No longer confined to assessing the condition of the state, attention was recalibrated to all fields of study, including medicine and sport, which is how we recognize statistics today. But while emerging fields like "machine learning" and "data mining" sound new and exciting now, "statistics" generally evokes memories of a dry and compulsory class taught in college or high school. In the book *Naked Statistics: Stripping the Dread from the Data,* author Charles Wheelan writes that students often complain that "statistics is confusing and irrelevant," but outside the classroom, they are glad to discuss batting averages, the wind chill factor, grade point averages and how to reliably measure the performance of an NFL quarterback.[7] As Wheelan observes, a large number of people study statistics as part of their education, but very few know how to apply these methods past examination day despite an inherent curiosity and interest in measuring things and especially performance.

This dichotomy has begun to change with the recent popularity of data science, which has grown in favor since Charles Wheelan's book was published in 2012. From the planning of data collection

---

[6] "Statistics," *Etymonline*, accessed July 2, 2017, www.etymonline.com/word/statistics/.
[7] Charles Wheelan, "Naked Statistics: Stripping the Dread from the Data," *W. W. Norton & Company*, First Edition, 2012.

to advanced techniques of predictive analysis, statistics is applied across nearly all corners of data science. Machine learning, in particular, overlaps with *inferential statistics*, which involves extracting a sample from a pool of data and making generalized predictions about the full population. Like inferential statistics, machine learning draws on a set of observations to discover underlying patterns that are used to form predictions.

In this absolute beginners' introduction to statistics, we focus primarily on inferential statistics to prepare you for further study in the field of data science and other areas of quantitative research where statistical inference is applied. While there is an excessive number of statistical methods to master, this introductory book covers core inferential techniques including hypothesis testing, linear regression analysis, confidence levels, probability theory, and data distribution. Descriptive methods such as central tendency measures and standard deviation are also covered in the first half of the book. These methods complement inferential analysis by allowing statisticians to familiarize themselves with the makeup and general features of the dataset. (In statistics, you can never be too familiar with your data.)

Before we proceed to the next chapter, it's important to note that there are four major categories of statistical measures used to describe data. Those four categories are:

1) Measures of Frequency: Analyzes the number of occurrences of any particular data value in the dataset and counts the number of times that it occurs, such as the number of Democrat and Republican voters within a sample population.

2) Measures of Central Tendency: Examines data values that accumulate in the middle of the dataset's distribution such as the median and mode. Discussed in Chapter 5.

3) Measures of Spread: Describes how similar or varied observed values are within the dataset such as standard deviation. Discussed in Chapter 6.

4) Measures of Position: Identifies the exact location of an observed value within the dataset such as standard scores. Discussed in Chapter 7.

2

STATISTICS 101

As a popular branch of mathematics, statistics involves studying how data is collected, organized, analyzed, interpreted, and presented. The goal of statistics is to determine the meaning of the data and whether variations, if any, are meaningful or due merely to chance. The alternative approach to statistical analysis is aptly named *non-statistical analysis* and is a form of qualitative analysis. Non-statistical analysis collects non-quantitative information gathered from text, sound, and moving and still images to inform decision-making. This form of analysis, however, is less scalable and practical for analyzing large trends and patterns. In this book, we will focus on quantitative analysis, and specifically, inferential statistics.

Descriptive vs Inferential Statistics

Quantitative analysis can be split into two major branches of statistics: *descriptive statistics* and *inferential statistics*. Descriptive statistics helps to organize data and provides a summary of data features numerically and/or graphically. Typical methods of descriptive statistics include the mode (most common value), mean (average value), standard deviation (variance), and quartiles. In general, descriptive statistics helps give you a better sense of your data and may be used in advance of inferential methods.

As a critical distinction from inferential statistics, descriptive statistical analysis applies to scenarios where all values in the dataset are known. In the case of an e-commerce website, this would mean access to information about all the registered users of the platform and then using descriptive statistics to summarize that information including the total average spending (mean), the most common bracket of spending (mode) or variance in customer spending activity (standard deviation). Using a complete dataset, the findings should accurately summarize and reflect the characteristics and patterns of the population. Sport, for example, draws heavily on descriptive statistics to generate individual player and team metrics, such as batting averages, assists, and points per game as each event is reliably logged and recorded.

In inferential statistics, there isn't the luxury of a full population. The analysis is instead subject to the nuances of probability theory, dealing with random phenomena, and inferring what is likely based on what is already known to be true.

When a full census isn't possible or feasible, a selected subset, called a *sample*, is extracted from the population. "Population" refers to an entire group of items, such as people, animals, transactions, or purchases. The selection of the sample data from the population is naturally subject to an element of randomness. Inferential statistics is then applied to develop models to extrapolate the study's findings from the sample data to draw inferences about the entire population while accounting for the influence of randomness.

To better explain how inferential statistics works, let's consider a large online platform such as YouTube, which as of 2017, entertains 1.5 billion logged-in users each month.[8] Rather than scrutinize the entire population of monthly logged-in users, we

---

[8] "YouTube has 1.5 billion logged-in monthly users watching a ton of mobile video," *TechCrunch*, accessed July 4, 2017, https://techcrunch.com/2017/06/22/youtube-has-1-5-billion-logged-in-monthly-users-watching-a-ton-of-mobile-video/.

can create a sample dataset of 10,000 users that can be analyzed to form predictions about the full population of YouTube users. Naturally, the 10,000 users we select for our analysis won't precisely represent and tell the story of 1.5 billion people. Although we can attempt to collect a sample dataset representative of the population, some margin of error is expected. In inferential statistics, we account for this margin of error with a statistical measure of prediction confidence called *confidence*.

Confidence is a measure to express how closely the sample results match the true value of the population. This takes the form of a percentage value between 0% and 100% called the *confidence level*. The closer the confidence level is to 100%, the more confident we are of the experiment's results fulfilling the true outcome. A confidence level of 95%, for example, means that if we repeat the experiment numerous times (under the same conditions), the results will match that of the full population in 95% of all possible cases. Alternatively, a confidence level of 0% expresses that we have no confidence in repeating the results in future experiments.

Note, also, that it's impossible to have a confidence level of 100%. The only way to prove the results are 100% accurate is by analyzing the entire population, which would render the study descriptive rather than inferential.

Hypothesis Testing

A crucial part of inferential statistics is the *hypothesis test,* in which you evaluate two mutually exclusive[9] statements to determine which statement is correct given the data presented. Due to the absence of a complete dataset, hypothesis testing is

---

[9] Mutually exclusive is a statistical term describing two or more events that cannot coincide or a situation where the occurrence of one outcome supersedes the other. Turning left and turning right, for example, are mutually exclusive because you can't do both at the same time. Tossing a coin and getting a heads or tails outcome is also mutually exclusive.

applied in inferential statistics to determine if there's reasonable evidence from the sample data to infer that a particular condition holds true of the population.

Hypothesis tests are constructed around a hypothesis statement, which is a prediction of a given outcome or assumption. You might hypothesize, for example, that a CEO's salary has no direct relationship with the number of syllables in their surname. In this example, the independent variable is the number of syllables in the CEO's surname and the dependent variable is the CEO's salary. If you discover that the independent variable does not affect the dependent variable, this outcome confirms the null hypothesis.

It's important to note that the term "null" does not mean "invalid" or associated with the value zero but rather a hypothesis that the researcher attempts or wishes to "nullify." Let's turn to an example to explain this concept further. For a long period of human history, most of the world thought swans were white, and black swans didn't exist inside the confines of mother nature. The null hypothesis that swans are white was later dispelled when Dutch explorers discovered black swans in Western Australia in 1697.

Prior to this discovery, "black swan" was a euphemism for "impossible" or "non-existent," but after this finding, it morphed into a term to express a perceived impossibility that might become an eventuality and therefore disproven. In recent times, the term "black swan" has been popularized by the literary work of Nassim Taleb to explain unforeseen events such as the invention of the Internet, World War I, and the breakup of the Soviet Union.

In hypothesis testing, the null hypothesis (H_0) is assumed to be the commonly accepted fact but that is simultaneously open to contrary arguments. If there is substantial evidence to the contrary and the null hypothesis is disproved or rejected, the *alternative hypothesis* is accepted to explain a given phenomenon. The alternative hypothesis is expressed as H_a or H_1.

Intuitively, "$_A$" represents "alternative." The alternative hypothesis covers all possible outcomes excluding the null hypothesis. To better conceptualize this relationship, imagine your niece or nephew challenges you to guess a number between 1 and 10, and you have one chance to guess the correct number. If you guess seven and that number is correct, then your null hypothesis stands as true. However, if you hypothesize any of the nine other possibilities as the null hypothesis and therefore guess incorrectly, you are effectively triggering the alternative hypothesis. Also, keep in mind, that the null hypothesis and alternative hypothesis are mutually exclusive, which means no result should satisfy both hypotheses.

Next, a hypothesis statement must be clear and simple. A clear hypothesis tests only one relationship and avoids conjunctions such as "and," "nor" and "or." According to California State University Bakersfield, a good hypothesis should include an "if" and "then" statement, such as: If [I study statistics] then [my employment opportunities increase].[10] The first half of this sentence structure generally contains an independent variable (i.e., study statistics) and a dependent variable in the second half (i.e., employment opportunities). A dependent variable represents what you're attempting to predict, and the independent variable is the variable that supposedly impacts the outcome of the dependent variable.

Hypotheses are also most effective when based on existing knowledge, intuition, or prior research. Hypothesis statements are seldom chosen at random.

Finally, a good hypothesis statement should be testable through an experiment, controlled test or observation. Designing an effective hypothesis test that reliably assesses your assumptions,

---

[10] "Formatting a testable hypothesis," *California State University Bakersfield*, accessed July 10, 2017,
http://www.csub.edu/~ddodenhoff/Bio100/Bio100sp04/formattingahypothesis.htm/.

however, is somewhat complicated and even when implemented correctly can lead to unintended consequences.

Cardiovascular disease, for instance, was a rare ailment until the second half of the 20th Century. After it became the leading cause of death in America in the 1950s, nutritional scientists began investigating possible causes for its increasing occurrence.[11] Their studies linked cardiovascular disease and high cholesterol levels with saturated animal fats contained in butter, meat, cheese, and eggs.

Later, in 1970, the *Seven Countries Study* released a comparative study of health and eating habits from a sample group of 12,700 middle-aged men in seven countries: Italy, Greece, Yugoslavia, Finland, the Netherlands, Japan, and the U.S.[12]

The study was carried out by the famous nutritionist Ancel Keys, who helped research and design the U.S. army's rations in World War II. The study found that men from countries consuming large amounts of saturated fats have a higher rate of heart disease than those living in countries that consume mainly grains, fish, nuts, and vegetables. Randomized controlled trials in the 1980's also concluded that saturated fats, including beef tallow, were unhealthy. These events and public protest provoked the fast-food giant McDonald's to modify their famous French fries recipe. More recent studies, though, have discovered that the proclaimed nutritional alternative to saturated animal fats, vegetable oils, are doing our body greater harm as a result of carboxylic acids.

Each of these studies started with a clear hypothesis, but they generated polarizing viewpoints regarding the question of whether we should consume food cooked in animal fats.

---

[11] Ancel Keys, "Home," *The Seven Counties Study*, accessed July 10, 2017, https://www.sevencountriesstudy.com/.
[12] Ancel Keys, "Home," *The Seven Counties Study*, accessed July 10, 2017, https://www.sevencountriesstudy.com/.

Research Design

Statistical-based research design can be separated into three categories: *experimental*, *quasi-experimental*, and *observational*. In an experimental study, subjects are randomly assigned to a test group, such as a control group and an experimental group. The control group is administered a standard form of medicine or even an absence of treatment, while the experimental group is assigned a new medicine or a form of treatment central to the study.

In the case of a quasi-experimental study, subjects are also assigned to an experimental, control or another group, but rather than relying on the research team's capacity to randomly assign subjects to test groups, the quasi-experimental method captures natural groups of test subjects, such as existing sports teams or university majors. Thus, rather than randomly splitting a football team into a control group and an experimental group, an entire football team can be assigned as the control group and another football team as the experimental group. This form of research is common in situations where it's impractical to conduct experimental design, such as field research where data is collected in a natural setting and conducted outside a laboratory environment.

For a design to be observational, the research team observes relationships and outcomes as they exist in the real world rather than relying on the assignment of separate test groups. This category of research is advisable in situations where the assignment of a control and experimental group qualifies as unethical and potentially harmful to participants, such as research on drink driving or drug abuse. Instead, research is constructed around analyzing subjects who already engage in the relevant behavior in a real-world setting.

The choice of experiment design will usually come down to practical and ethical concerns as well as financial considerations.

3

DATA SAMPLING

For most large-scale studies, it's prohibitively expensive or impossible to survey and analyze an entire population of items, people, objects, etc. Instead, statisticians typically gather a sample, which is a subset of data that is representative of the full population. This, as we know, belongs to the field of inferential statistics.

For the statistician to be confident of the experiment's results, the sample needs to be a reliable representation of the population data. This is to ensure that inferences and conclusions drawn from the sample data reliably reflect overall trends.

To collect sample data, statisticians use a range of techniques based on random sampling. For the data to be reliable, each individual sample (item, person, object, etc.) must have an equal probability of selection and measurement. While simple, in theory, this can be difficult to implement and especially for volunteer-based surveys. Suppose, for example, you send a survey to customers regarding your recent refer-a-friend promotion (i.e. refer a friend to earn free credit). You send the survey to your entire email list without filtering for age, gender, and country. The survey might appear representative of your population because each survey recipient was allocated an equal opportunity to respond. Unfortunately, however, the survey results aren't a representative cross-section of your customer base. First, customers who received financial commissions from the promotion are over-represented in the survey results because they have a vested interest in the activity and want to see the program continued. Second, customers who received the email during work hours might be more likely to view and answer the survey from their work station over customers who received the survey outside work hours in another time zone. As not everyone's opinion has an equal chance of being measured, this could lead to a potential sampling bias.

A similar problem occurred during the 2012 U.S. Presidential Election when the Gallop Poll's reliance on sampling using landline telephones led to a severe underrepresentation of younger voters who were unavailable for contact by telephone. Due to flaws in their sampling procedures (which had proved accurate over many decades), the poll failed to derive a representative sample in the smartphone era and the Gallop Poll mispredicted the eventual outcome of the election.

Another potential problem with respondent sampling occurs when researchers disclose the topic of the survey to respondents in advance. Sensitive topics such as tax fraud and extramarital affairs may lead to a high dropout rate of responses from relevant offenders. For this reason, many surveys don't indicate ahead of time the topic of the questions or they disguise the hypothesis question with irrelevant questions they're not interested in analyzing.[13]

Independent Data

Next, each data observation must be independent. This means that the selection of one item should not impact the selection of another item. If we wish to measure public sentiment, we need to ask: will respondents be influenced by the remarks of a fellow respondent who is a well-respected celebrity that makes their comments public before the survey? If so, this is not a case of independent sample data but rather dependent sample data.

---

[13] Daniel Levitin, "A Field Guide to Lies and Statistics," *Penguin*, 2016.

Group surveys, in particular, are susceptible to interference that can affect the independence of individual responses due to social forces such as peer pressure, agreeing with the first person to speak or a proclivity to avoid conflict and maintain group cohesion.

Information Censoring

When a significant drop-out occurs among participants (especially long-term studies) this can incite biased results as variance within the sample is absent from the final sample. This problem is particularly pronounced in medical or nutritional experiments where respondents experiencing adverse side-effects withdraw before the experiment has concluded.

Social Desirability Bias

Social desirability generates bias when respondents are inclined to present themselves in a favorable light or respond in a way that aligns with desirable social norms. This type of bias is most common in group surveys and one-on-one conversations with an interviewer and can also occur in anonymous written surveys.

Random Selection

It's important to randomly allocate respondents to groups, such as the control group and experimental group, so that each respondent has an equal chance of selection—similar to a lottery draw. In addition, you can go a step further by incorporating a double-blind, where both the participants and the experimental team aren't aware of who is allocated to the experimental group and the control group respectively.

Collection Bias

Lastly, each data observation must be unbiased and unmanipulated by the collecting party. A high-profile example of this type of data bias appeared in Google's "Flu Trends" program.

The program, which started in 2008, intended to leverage online searches and user location monitoring to pinpoint regional flu outbreaks. Google collected and used this information to tip-off and alert health authorities in regions they identified. Over time the project failed to accurately predict flu cases due to changes in Google's search engine algorithm. A new algorithm update in 2012 caused Google's search engine to suggest a medical diagnosis when users searched for the terms "cough" and "fever." Google, therefore, inserted a false bias into its results by prompting users with a cough or a fever to search for flu-related results (equivalent to a research assistant lingering over respondents' shoulder whispering to check the "flu" box to explain their symptoms). This increased the volume of searches for flu-related terms and led Google to predict an exaggerated flu outbreak twice as severe as public health officials anticipated.[14]

The inability to draw a perfectly random sample from the subject population is a common theme in data collection from the real world. Avoiding obvious pitfalls, such as collection bias, helps to preserve the integrity of your data. Equally, it's important to put your findings into perspective by sharing with your audience where and how your data was obtained as well as the potential limitations.

How Much Sample Data is Needed?

Now that we've established the need for sample data to be independent, randomized, and unbiased, the next question is how much sample data do we actually need? This is not a straightforward question to answer, as the optimal amount depends on the goals of the analysis.

The first tip to remember is that more relevant sample data is usually better than less. While more data allows you to cover

---

[14] William Isaac & Andi Dixon, "Why Big-Data Analysis of Police Activity Is Inherently Biased," *Observer*, accessed July 18, 2017, http://observer.com/2017/05/why-big-data-analysis-of-police-activity-is-inherently-biased/.

more potential combinations and generally leads to more accurate predictions, there is no hard-and-fast rule on how much data is needed to perform statistical analysis. In some cases, it might not be possible or cost-efficient to source data for every possible combination, and, in which case, you have to make do with the data you have available.

Secondly, statistical analysis performs best when the dataset contains a full range of feature combinations. What does a full range of feature combinations look like? Imagine a dataset about statisticians that possess the following features:

- University degree
- Online certification
- Children
- Salary

To analyze the relationship that these three features have to a statistician's salary, we need a dataset that records the value of each combination of features. For instance, we need to know the salary for a statistician with a university degree, online certification, and who has <u>no</u> children, as well as a statistician with a university degree, online certification, and who <u>has</u> children.

The more combinations, the more effective the analysis is at capturing how each attribute affects the statistician's salary.

4

ODDS & PROBABILITY

As mentioned in Chapter 2, we can't prove a hypothesis made about the future using inferential statistics. The best we can do is determine how likely or probably something is, which leads us to the sub-field of probability.

Probability is the likelihood of something happening and is typically expressed as a number with a decimal value called a floating-point number. The closer the floating-point number is to 1.0, the more likely the hypothesis is true.[15] If $P(A) = 0.02$, the probability of "A" occurring is 2%. This means that A is not very likely to occur. If $P(A) = 0.99$, A is considered almost a certainty. As seen in this example, percentages can also be used instead of decimal placing, i.e., 0.5 or 50%.

Probability, however, is not the same as odds. While probability expresses the likelihood of something happening expressed in percentage or decimal form, odds define the likelihood of an event occurring with respect to the number of occasions it does not occur. For instance, the odds of selecting an ace of spades from a standard deck of 52 cards is 1 against 51. On 51 occasions a card other than the ace of spades will be selected from the deck.

Another related concept to probability is correlation[16] which is often computed during the exploratory stage of analysis to

---

[15] In statistics, you can specify probability using decimals within the range of 0.0 and 1.0 or as percentages within 0% and 100%.

understand general relationships between variables. Correlation describes the tendency of change in one variable to reflect a change in another. For instance, a person's way of speaking and accent is usually highly correlated with the people that live in their proximity. Correlation is also different from *causation*, which captures a causal relationship between variables. This means that just because two items reflect the same pattern, correlation alone doesn't represent that changes in one item affect the other. This concept that an observed correlation cannot explicitly prove causation is captured in the famous slogan "correlation does not imply causation."

Correlation and causation, though, are commonly confused and for obvious reasons; when two variables vary together, they can appear connected. But the observed correlation could be caused by a third and previously unconsidered variable, known as a *lurking variable* or *confounding variable*. As an example, the variable "level of sunlight on weekdays" might appear strongly correlated with people waking up early in the morning but that variable is an unlikely cause. There is an active lurking variable, which is the reality that most people wake up early to go to work or attend class on those days of the week. It's important to consider variables that fall outside your hypothesis test as you prepare your research and before publishing your results.

Another case of confusing correlation and causation arises when you analyze too many variables while looking for a match. This is known as the *curse of dimensionality*. (In statistics, dimensions can also be referred to as variables. So, if we are analyzing three variables, the results fall into a three-dimensional space.) You can find instances of the "curse" or phenomenon using Google Correlate (www.google.com/trends/correlate). When you enter a

---

[16] ·"Correlation" is also commonly referred to as the "correlation coefficient" or even the "Pearson correlation," ·which is the most common measure of correlation. Expressed as a number between -1 and 1, the Pearson correlation is a measure of linear association between two variables. A perfect positive linear relationship is equal to 1 and 0 indicates no relationship.

query into the search box, Google will find search terms that share a similar pattern of historical search activity. If I enter my employer's name, Alibaba Cloud, for example, I receive the following results as shown in Figure 1.

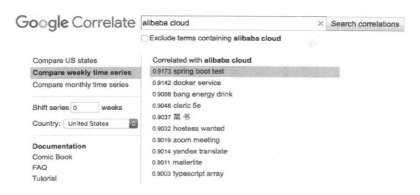

Figure 1: Google Correlate results for "alibaba cloud"

After matching search patterns for the keyword "alibaba cloud," Google retrieves other search terms that are highly correlated in terms of historical search patterns (i.e. volume, duration, peaks and troughs, etc.). The first two results "spring boot test" (0.9173) and "docker service" (0.9142) are related to Alibaba Cloud's product offering and are likely to be causal in nature. Other results including "bang energy drink" (0.9086) and "hostess wanted" (0.9032) seem less plausible relationships.

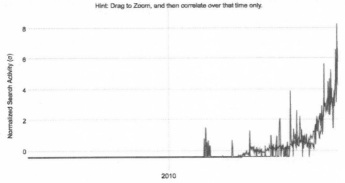

Figure 2: Google Correlate's comparison of "alibaba cloud" & "bang energy drink"

It turns out that the bang energy drink came onto the market at a similar time as Alibaba Cloud's international product offering and then grew at a similar pace in terms of Google search volume. It's difficult to imagine, though, that consumers of the bang energy drink have an affinity for Alibaba Cloud to satisfy their IT hosting and big data needs. The danger with Google Correlate (which can be precise too with their results) is that with millions of variables at play, it's bound to find at least a handful of variables that follow similar trends purely by coincidence.

Note that the curse of dimensionality tends to affect machine learning and data mining analysis more than traditional hypothesis testing due to the high number of variables placed under consideration.

Independent Events vs Conditional Probability

In statistics, we're interested in relationships between variables and whether their apparent covariance is meaningful. Those variables can either be independent or conditional. In probability, two events are considered independent if the occurrence of one event does not influence the outcome of another event. This means that the outcome of one event, such as flipping a coin, doesn't predict the outcome of another. If you flip a coin twice, the outcome of the first flip has no bearing on the outcome of the second flip.

To draw another example, imagine it's raining outside. This fact does not affect the size of your house. It does, however, affect whether the lawn is damp or dry. Based on this example, rain and house size would be independent whereas rain and lawn are dependent. The occurrence of one variable, the rain, affects the probable occurrence of the other, damp lawn. Calculating probabilities when two variables are conditional is somewhat different from when they are independent of one another.

The probability of one event (E) given the occurrence of another conditional event (F) is expressed as P(E|F), which is the probability of E given F. A damp lawn (E) can therefore be expressed as P(E|F), which is the probability of dampness based on whether it has rained (F).

Conversely, two events are said to be independent if P(E|F) = P(E). This equation holds that the probability of E is the same irrespective of F being present.

P(damp lawn |rain) = P(damp lawn)

In this example, the lawn was already damp because you watered the lawn in advance of the unexpected storm. This expression can also be tweaked to compare two sets of results where the conditional event (F) is absent from the second trial.

P (grade|energy drink) > P (grade|no energy drink)
620/1000 > 500/1000

This example expresses that students' exam performance is conditional on the consumption of an energy drink. One of the most widely used applications of conditional probability is based on Bayes' theorem, as we'll explore in the next section.

Bayes Theory

Thomas Bayes, an English Presbyterian minister who moonlighted as a statistician and a philosopher, explored the concept of conditional probability in the middle of the 18th Century. In the book *Antifragile: Things That Gain from Disorder,* author Nassim Taleb notes (albeit satirically) the high concentration of discoveries that originated from British clergyman, which includes the power loom, the terrier, the planet Uranus, modern archaeology, the Malthusian Trap, and of course, Bayes' theorem.

> "An extraordinary proportion of work came out of the rector, the English parish priest with no worries, erudition, a large or at least comfortable house, domestic help, a reliable supply of tea and scones with clotted cream, and an abundance of free time."[17]

It's actually thought that Bayes learned mathematics and probability from a book by Abraham de Moivre, who is introduced in Chapter 6.[18] Like Gerolamo Cardano, Reverend Thomas Bayes' manuscript on the topic of conditional probability was published posthumously. The manuscript, *An Essay towards solving a Problem in the Doctrine of Chances*, was edited and read by his friend Richard Price at the English Royal Society in 1763, two years after Thomas Bayes' death in 1761.

The theory was developed further by the Frenchman Pierre-Simon Laplace, who in 1812 wrote *Théorie analytique des probabilités.* However, it wasn't until 1939, when Sir Harold Jeffrey and Bertha Swirles published *Theory of Probability*, that the Bayesian view of probability started to receive widespread attention.

---

[17] Nassim Taleb, "Antifragile: Things That Gain From Disorder," *Random House,* 2012.
[18] George Barnard, "Thomas Bayes—a biographical note," *Biometrika*, 1958.

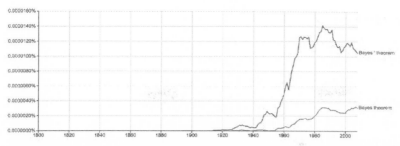

Figure 3: Historical mentions of Bayes/Bayes' theorem in published books, Source: Google Ngram Viewer

British statistician and mathematician Sir Harold Jeffrey later wrote that Bayes' theorem "is to the theory of probability what the Pythagorean theorem is to geometry."[19] The theory has also experienced a renewed interest in recent years following statistician Nate Silver's successful forecast of the 2012 U.S. presidential election.

The premise of this theory is to find the probability of an event, based on prior knowledge of conditions potentially related to the event. For instance, if reading books is related to a person's income level, then, using Bayes' theory, we can assess the probability that a person enjoys reading books based on prior knowledge of their income level. In the case of the 2012 U.S. election, Nate Silver drew from voter polls as prior knowledge to refine his predictions of which candidate would win in each state. Using this method, he was able to successfully predict the outcome of the presidential election vote in all 50 states.

Whether you're interested in studying political outcomes, animal populations, or the risks involved in lending money to a potential borrower, Bayes' theorem provides a powerful tool in probability and inferential statistics.

---

[19] Harold Jeffreys, "Scientific Inference," *Cambridge University Press,* Third Edition, 1973, p. 31.

Learning the Formula

In the biography of American physicist Richard Feynman, *Surely You're Joking, Mr. Feynman!,* the author shares a valuable lesson about science that holds true of statistics and probability.

In an anecdote from the book, Professor Feynman argues against disseminating definitions in favor of urging students to learn through practical application. While reviewing a Brazilian physics textbook, Feynman flicks to a page containing a definition for "triboluminescence" and says to his audience:

"Triboluminescence. Triboluminescence is the light emitted when crystals are crushed..."

..."And there, have you got science? No! You have only told what a word means in terms of other words. You haven't told anything about nature–what crystals produce light when you crush them, why they produce light. Did you see any student go home and try it? He can't."

"But if, instead, you were to write, 'When you take a lump of sugar and crush it with a pair of pliers in the dark, you can see a bluish flash. Some other crystals do that too. Nobody knows why. The phenomenon is called 'triboluminescence.' Then someone will go home and try it. Then there's an experience of nature."

Like triboluminescence, you can stare for hours at the formula for Bayes' theorem and feel lost and confused. The theory, though, is relatively simple once applied and put into practice.

$$P(A|B) = \frac{P(A)\ P(B|A)}{P(B)}$$

As discussed, the conditional probability is the probability that something will happen, *given that something else* has happened. Using an example, let's suppose on any given day the probability of the lawn being damp is 5%. However, if we know that it's winter, that probability is assumed to be higher. If we refer to lawn dampness as event A and winter as event B, then this relationship would be stated as the probability of A (lawn dampness) given B (it's winter), or P(A|B).

The calculation of conditional probabilities is laid out by Bayes' theorem. This mathematical formula, as shown again below, can be used to predict how updated information about one variable can help us to predict changes in the probability of another variable occurring. This formula can also be used in statistical inference to make inferences about the entire population.

$$P(A|B) = \frac{P(A)\ P(B|A)}{P(B)}$$

In this equation, "A" and "B" represent two events, while "P" represents the probability of that event occurring. Thus:

P(A|B) is the probability of A given that B happens (conditional probability)

P(A) is the probability of A without any regard to whether event B has occurred (marginal probability)

P(B|A) is the probability of B given that A happens (conditional probability)

P(B) is the probability of B without any regard to whether event A has occurred (marginal probability)

Both **P(A|B)** and **P(B|A)** are the conditional probability of observing one event given the occurrence of the other. Both **P(A)** and **P(B)** are marginal probabilities, which is the probability of a variable without reference to the values of other variables.

As mentioned, the equation is difficult to understand without a practical demonstration. Let's imagine a particular drug test is 99% accurate at detecting a subject as a drug user. Suppose now that

5% of the population has consumed a banned drug. How can Bayes' theorem be applied to determine the probability that an individual, who has been selected at random from the population is a drug user if they test positive?

If you designate event A as "drug user" and event B as "testing positive for the drug," it fits into the Bayes' formula as follows:

$$P(A|B) = \frac{P(A)\ P(B|A)}{P(B)}$$

P(user | positive test) = P(user) * P(positive test | user) / P(positive test)

P(A|B) is the probability of a drug user given a positive test result

P(A) is the probability of being a drug user without any regard to the result of the test

P(B|A) is the probability of a positive test result given that the individual is a drug user

P(B) is the probability of positive test result without any regard to whether the individual is a drug user

We already know that P(user) is equal to 0.05 and P(positive test | user) is equal to 0.99. However, the probability of returning a positive test regardless of whether someone is a user, **P(positive test)**, still needs to be calculated.

$$P(\text{user} \mid \text{positive test}) = \frac{(0.05)\ (0.99)}{P(\text{positive test})}$$

There are two elements to P(positive test): someone that tests positive and who is a user, and someone that tests positive who is not a user (false-positive). This expands P(positive test) to the following equation:

P(positive test) = P(positive test | user) * P(user) + P(positive test | non-user) * P(non-user)

P(positive test) = (0.99 * 0.05 + 0.01 * 0.95)

Let's now plug in P(positive test)

$$P(user \mid positive\ test) = \frac{(0.05)\,(0.99)}{0.99*0.05 + 0.01*0.95}$$

(0.05 * 0.99) / (0.99 * 0.05 + 0.01 * 0.95)

0.0495 / (0.0495 + 0.0095)

0.0495 / 0.059 = 0.8389

Using Bayes' theorem, we're able to determine that there's an 83.9% probability that an individual with a positive test result is an actual drug user. The reason this prediction is lower for the general population than the successful detection rate of actual drug users or P(positive test | user), which was 99%, is due to the occurrence of false-positive results.

For example, if we test 10,000 individuals and 95% are non-users (9,500 people), then the prediction error of 1% produces 9,500 * 0.01 = 95 false-positive results. Therefore, of the individuals who test positive, 95 are non-users who test positive and 495 (500 * 0.99) are actual drug users. In total, 590 (95 + 495) people test positive, of which:

495 / 590 = 83.89% are drug users

95 / 590 = 16.11% are non-drug users (false-positive)

	Tested Positive	Tested Negative	Total
Drug User	495	5	500
Non-User	95	9405	9500
Total	590	9410	**10000**

Table 1: Summary of test results

For those who test negative, 9410 people test negative, of which:

5 / 9410 = 0.053% are drug users (false-negative)

9405 / 9410 = 99.947% are not drug users

In this example, P(B) or the probability of a positive test result for all individuals was unknown, and this is characteristic of medical test examples. For other Bayes' examples, you will often find that P(B) is already provided. You will also notice that Bayes' theorem can be written in multiple formats including the use of ∩ (*intersection*).

$$P(A|B) = \frac{P(A) \, P(B \cap A)}{P(B)}$$

Finally, it's important to acknowledge that Bayes' theorem can be a weak predictor in the case of poor data regarding prior knowledge and this should be taken into consideration.

Binomial Probability

Another useful area of probability is binomial probability, which is used for interpreting scenarios with two possible outcomes. Pregnancy and drug tests both produce binomial outcomes in the form of negative and positive results, as does flipping a two-sided coin.

The probability of success in a binomial experiment is expressed as **p**, and the number of trials is referred to as **n**. If we flip 100 coins, n is equal to 100, and the successful outcome of a head is p 0.5 (50%). This is easy to understand and to test. But what about drawing aggregated conclusions from multiple binomial experiments such as flipping consecutive heads using a fair coin? Here, you would need to calculate the likelihood of multiple independent events happening, which is the product (multiplication) of their individual probabilities.

Given that we know the probability of flipping heads with a single coin is 0.5, we can multiply the individual probability of two flips (0.5 x 0.5) to produce the final probability (0.25).

Next, let's consider an example with three binomial tests. Suppose we invite three friends to dinner, and we predict how many friends accept the invitation. For each person, there are two possible outcomes: a positive outcome (yes) and a negative outcome (no), and each outcome is independent of the response of the other two people invited.

The dinner, by the way, is all expenses paid but limited to you and one guest. In addition, it cannot be canceled or postponed. You decide to invite all three friends because you have a hunch that two friends will opt out. Remember, you can only take one friend, and ideally, you want to avoid canceling on any of your friends. At the same time, you want to avoid three friends declining and spending the night alone.

You make some estimates regarding the probability of your friends accepting your invite.

Based on your recently estranged friendship with Alex you log him as p 0.30 (30%).

Based on previous attendance you log Casey as p 0.60 (60%).

Finally, you concede there's a p 0.10 (10%) probability of River agreeing to attend given her hectic weekend schedule.

Let's first multiply the chances of Alex saying "yes" and Casey and River both declining with a "no."

Name	Yes	No	Accepts
Alex	0.3	0.7	Yes
Casey	0.6	0.4	No
River	0.1	0.9	No

Table 2: Scenario 1 of only Alex attending

This can be calculated as: **0.3 x 0.4 x 0.9 = 0.108 (10.8%)**

There is a 10.8% probability that Alex accepts your invitation and that both Casey and River decline.

It's time now to formulate a prediction for Casey.

Name	Yes	No	Accepts
Alex	0.3	0.7	No
Casey	0.6	0.4	Yes
River	0.1	0.9	No

Table 3: Scenario 2 of only Casey attending

0.7 x 0.6 x 0.9 = 0.378 (37.8%)

There is a 37.8% probability of Casey joining you on Saturday night and Alex and River declining. Lastly, let's look at the prediction for River.

Name	Yes	No	Accepts
Alex	0.3	0.7	No
Casey	0.6	0.4	No
River	0.1	0.9	Yes

Table 4: Scenario 3 of only River attending

0.7 x 0.4 x 0.1 = 0.028 (2.8%)

No need to get your hopes up but River is still a possibility at 2.8%.

After finding the probability of three individual outcomes, we can now determine whether the hypothesis of a single person saying "yes" is valid. The answer is Alex **0.108** + Casey **0.378** + River **0.028** = **51.4** (51.4%).

Although it's a delicate way of managing your social life, you theoretically have a higher chance of succeeding in this experiment (p 0.514) than flipping heads (p 0.5) with a two-sided coin.

You may also be wondering the probability of no one accepting your invitation or the possibility of all three friends responding "yes." A full sequence of possibilities is listed in the following table:

Friends That Accept	Calculations	Outcome
0	0.7 x 0.4 x 0.9	0.252 (25.2%)
1	0.108 + 0.378 + 0.028	0.514 (51.4%)
2	(0.3 x 0.4 x 0.1) + (0.3 x 0.6 x 0.9) + (0.6 x 0.1 x 0.7)	0.216 (21.6%)
3	0.3 x 0.6 x 0.1	0.018 (1.8%)
		1.0 (100%)

Table 5: All possible combinations

The possibility of no one accepting is 25.2%, two friends accepting is 21.6%, and all three friends accepting is 1.8%.

Permutations

Permutations are another useful tool to assess the likelihood of an outcome. While it's not a direct metric of probability, permutations can be calculated to understand the total number of possible outcomes, which can be used for defining odds.

Permutations are all around us. On any given day we are faced with the challenge of choosing from a daunting range of possibilities.

I've forgotten the four-digit code to my bike or Airbnb apartment, but I remember the first digit, what is the maximum number of combinations I need to test in order to find those three missing digits?

One way of dealing with the dread of combinations is to calculate the full number of permutations, which refers to the maximum number of possible outcomes from arranging multiple items. For instance, there are only two possible outcomes from flipping a single coin or seating two people at a table with two seats. In the case of the latter, one person can sit by the window, and the other person can sit by the door and vice versa. However, if we add a third chair and a third person, suddenly there are six possible combinations for where three people can sit.

To find the full number of seating combinations for a table of three, we can apply the function three-factorial, which entails multiplying the total number of items by each discrete value below that number, i.e., 3 x 2 x 1 = 6. Yes, it's that simple. Want to add a fourth guest? Four-factorial is 4 x 3 x 2 x 1 = 24. To seat four people around a table, there is a total of 24 combinations.

Another way of using permutations is for horse betting. Let's say you want to know the full number of combinations for randomly picking a box trifecta, which is a scenario where you select three horses to fill the first three finishers in any order.

This exercise requires a variation of the calculation used in the previous example. For this task, we're not just calculating the total number of permutations but also a subset of desired possibilities (recording a 1st place, recording a 2nd place, and recording a 3rd place finish).

Let's calculate the total possible combinations that lead to a top three placing among a field of 20 horses. The total number of combinations on where each horse can finish is calculated as 20 x 19 x 18 x 17 x 16 x 15 x 14 x 13 x 12 x 11 x 10 x 9 x 8 x 7 x 6 x 5 x 4 x 3 x 2 x 1.

Twenty-factorial = 2,432,902,008,176,640,000

We next need to divide twenty-factorial by seventeen-factorial to ascertain all possible combinations of a top three placing.

17 factorial is calculated as 17 x 16 x 15 x 14 x 13 x 12 x 11 x 10 x 9 x 8 x 7 x 6 x 5 x 4 x 3 x 2 x 1.

Seventeen-factorial = 355,687,428,096,000

Twenty-factorial / Seventeen-factorial = 6,840

Thus, there are 6,840 possible combinations among a 20-horse field that will offer you a box trifecta. The fact that there are 2,432,902,008,176,633,160 other combinations where you won't win a box trifecta might also make you think twice before making that bet! (This outcome, though, is based on randomly drawing three horses to complete a box trifecta and does not consider unique knowledge of each horse. It's, therefore, possible to optimize your bet using domain knowledge.)

5

CENTRAL TENDENCY

With data all around us, we need a way to make sense of it all. This can take the form of advanced analysis, but it can also start with simple techniques that identify the central point of a given dataset, known as *central tendency measures*. The three primary measures of central tendency are the *mean*, *mode*, and *median*. Although sometimes used interchangeably by mistake, each of these terms has a distinct meaning and method for identifying the central point of a given dataset. This chapter also gives inclusion to a fourth measure called the *weighted mean*.

While the following methods are examples of descriptive statistics, it's common practice to use these techniques in preparation for inferential analysis as a way of familiarizing yourself with the dataset.

The Mean

The mean, the term most commonly used to describe the midpoint of a dataset, is the average of a set of values and ·the easiest central tendency measure to understand. The mean is calculated by dividing the sum of all numeric values by the number of observations contained in that sum.

I.E., The mean for 2 + 4 + 1 + 1 is:

8 / 4 observations = 2

Although it is useful for describing a dataset, the mean can be highly sensitive to outliers. To offset this problem, statisticians sometimes use the *trimmed mean*, which is the mean obtained after removing extreme values at both the high and low band of the dataset, such as removing the bottom and top 2% of salary earners in a national income survey.

The Median

The next measure of central tendency is the median. Rather than aggregate all observations into an average value, as you would with the mean, the median pinpoints the data point(s) located in the middle of the dataset to suggest a viable midpoint. The median, therefore, occurs at the position in which exactly half of the data values are above and half are below when arranged in ascending or descending order.

To better understand this method, let's review the following dataset, which logs driving license test scores for 13 drivers sequenced from highest to lowest.

90 89 84 80 77 70 *70* 59 55 33 32 31 30

In this small dataset, the median value is 70—not because that's the average value of the dataset but because that's the value positioned in the middle of the dataset. As there are six other data points on either side, 70 is the median value.

What, though, if there's an even number of data points that can't be split cleanly as in the previous example? The solution for an even number of data points is to calculate the average of the two middle points as shown here:

90 89 84 80 77 *70 59* 55 33 32 31 30

The two middle data points have exactly five data points on either side. Their average is then calculated as (70 + 59) / 2. Thus, the median for the dataset is 64.5.

Finding the mean and the median can be useful techniques for asking questions and challenging the accuracy of statistical-based claims you see reported in the media, advertising, and other studies. At times it may be convenient for a politician to reference the mean value of middle-class income but better for his or her political rival to highlight the median value.

The mean and median sometimes produce similar results, but, in general, the median is a better measure of central tendency than the mean for data that is asymmetrical as it is less susceptible to outliers and anomalies. As an example, consider the original dataset with one change:

900 89 84 80 77 ***70 59*** 55 33 32 31 30

The mean for this dataset is now 128.33, while the median is still 64.5. Here, the median is unaffected by the presence of the first data point (900) which is greater in value than the other data points combined. The median is thus a more reliable metric for skewed (asymmetric) data or discrediting the old joke about the average net worth of a room of people in the company of Bill Gates.

The Mode

The next statistical technique to measure central tendency is the *mode*, which is an even more straightforward technique than the mean and median. The mode is the data point in the dataset that occurs most frequently. It's generally easy to locate in datasets with a low number of discrete categorical values (a variable that can only accept a finite number of values) or ordinal values (the categorization of values in a clear sequence) such as a 1 to 5-star rating system on Amazon. In the case of an Amazon product rating, there are five available values: 1, 2, 3, 4, and 5.

Figure 4: Customer reviews of a book sold on Amazon.com

In Figure 4, the mode is 5-stars as that's the single most common value in the dataset (47%), compared to the next most common data point of 4-stars (23%). The effectiveness of the mode can be arbitrary and depends heavily on the composition of the data. The mode, for instance, can be a poor predictor for datasets that do not have a single high number of common discrete outcomes. For example, is selecting the mode (5-stars) for the following product rating a true reflection of the dataset's midpoint?

Figure 5: Customer reviews of another book sold on Amazon.com

Weighted Mean

The final statistical measure of central tendency factors the *weight* of each data point to analyze the mean. A typical example of this method can be seen in student assessment.

Over the course of a semester, students' grades are split into four assessment pieces, with the final exam accounting for 70% of the total grade. Weights are then added by pairing each assessment piece with its overall contribution to the final grade, expressed in proportion to the value of total assessment.

Assessment	Weight	Result	Weighted Result
Test 1:	0.1	70/100	0.07
Test 2:	0.1	50/100	0.05
Test 3:	0.1	90/100	0.09
Final Exam:	0.7	80/100	0.56
Overall Grade			0.77

Table 6: Sample dataset of student assessment

The final grade is 77 on a scale of 100. This particular case demonstrates that although the student struggled at the start of the semester, their first two results had a modest impact on their final grade.

What, though, if the student was graded based on the mean and median respectively?

The **mean** = **72.5**   (70 + 50 + 90 + 80 = 290/4)

The **median** = **70**   (70 + *50 + 90* + 80 = 140/2)

The **mode,** in this case, cannot be identified, as there are no common outcomes but rather four single discrete values.

As demonstrated in this chapter, a suitable measure of central tendency depends on the composition of the data. The mode, for example, is easy to locate in datasets with a low number of discrete values or ordinal values, whereas the mean and median are suitable for datasets that contain continuous variables. Finally, the weighted mean is used when you want to emphasize a particular segment of data without disregarding the rest of the dataset.

6

MEASURES OF SPREAD

Despite pre-flight defect detection and careful engineering, the aviation industry was struggling to keep jet-powered planes in the air in the late 1940s. With World War II over, the number of military plane crashes was as many as seventeen in a single day.[20]

Senior U.S. Air Force officials blamed the pilots for the "error." But the pilots weren't ready to accept responsibility for the crisis. After multiple official inquiries culminated in no new evidence, officials commissioned an internal audit in 1950 of 4,063 pilots at Wright Air Force Base in Ohio. This followed internal concerns that the physical stature of pilots had changed since the original design of the cockpit in 1926.

The official audit measured pilots' body dimensions and was based on 140 different measurements, including thumb length and distance of the eye to the ear. The Air Force believed that an updated assessment of the pilots' body size might bring forward an argument to remodel the cockpit. Their hypothesis, though, was rejected by Lt Gilbert S. Daniels, a 23-year-old scientist who, until recently, had never set foot on an airplane.

Daniels had majored in physical anthropology at Harvard University where he ran experiments analyzing the shape and size of students' hands as part of his undergraduate thesis. His college thesis was noteworthy not because it type-casted the

---

[20] Todd Rose, "The End of Average: How We Succeed in a World That Values Sameness," *HarperCollins Publishers Ltd, 2016.*

hand-size of the average American male but because the so-called average failed to resemble the measurement of any one individual in the study. It was Daniels' keen understanding of body averages that ultimately led him to challenge his research colleagues at Aero Medical Laboratory.

Daniels injected a fresh perspective, but as a junior researcher, his initial involvement was limited to tape measure duties. Daniels, though, later utilized data gathered from the Air Force study to examine the 10 physical dimensions deemed most relevant for cockpit design. Using this shortlist of relevant measurements, he attempted to calculate and generate a more realistic audit of the average pilot. Rather than using the mean, Daniels padded the average by taking the middle 30 percent of measurements for each dimension. Daniels then compared the average size of the 10 dimensions against each pilot.

He soon found that of the 4,063 men involved in the study, not one of the pilots matched with the average range for all 10 dimensions. His findings uncovered that one pilot might qualify as average based on arm span and leg length but at the same time record above or below the average based on another physical measurement, thereby disqualifying him from any unified average.

Daniels scaled his analysis back to three dimensions, but, even then, less than 3.5 percent of pilots met the requirements of "average." The results were all the more unusual given the military's controlled hiring policy that purged applicants on the extreme end of physical proportions—making the sample a far cry from the ragtag sample of students he assessed on campus in Boston.

The results made Daniels' suspicions clear: the average pilot didn't exist and designing the cockpit to fit central tendency measures based on the "average pilot" was designing the cockpit for no one.[21] His findings would eventually spark a redesign of

---

[21] Todd Rose, "The End of Average: How We Succeed in a World That Values

the cockpit that included customized seat fitting and other elements of personalization to fit the individual pilot.

Daniels' study illustrates how variability can manipulate the reliability of the mean and other central tendency measures discussed in the previous chapter. This also brings us to the important topic of statistical measures of spread, which describes how data varies. As the recordings of limbs, events, objects, and other items change, it's important to analyze and acknowledge this variance when making decisions.

To explore this phenomenon, let's take a closer look at data variability using the following two datasets.

Dataset 1	Dataset 2
2	1
2	2
10	1
10	20
Mean: 6	Mean: 6
Range: 8	Range: 19

Table 7: Sample datasets with a mean of 6

While these two datasets share the same mean (6), notice how the values in Dataset 1 vary from those in Dataset 2. The composition of the two datasets is different despite the fact they each dataset has a mean of 6. The critical point of difference is the *range* of the datasets, which is a simple measurement of data variance. As the difference between the highest value (maximum) and the lowest value (minimum), the range is calculated by subtracting the minimum from the maximum. For

Sameness," *HarperCollins Publishers Ltd, 2016.*

Dataset 1, the range is "8" (10 − 2), and for Dataset 2 it is "19" (20 − 1).

A weakness of this measure is that one data point can greatly impact the results. As seen in Dataset 2, the last value of "20" exaggerates the low range between the other three data points, which otherwise have a combined range of "1" (2 − 1). In this way, it only takes one abnormal data point to inflate the range of the dataset.

On the other hand, knowing the range for the dataset can be useful for data screening and identifying errors. An extreme minimum or maximum value, for example, might indicate a data entry error, such as the inclusion of a measurement in meters in the same column as other measurements expressed in kilometers.

Standard Deviation

While measuring the range of the dataset can be useful, this technique, as we know, is susceptible to anomalies. A practical alternative for interpreting data variability is *standard deviation*, which describes the extent to which individual observations differ from the mean. Expressed differently, the standard deviation is a measure of the spread or dispersion among data points[22] and is just as important as central tendency measures for understanding the underlying shape of the data.[23]

Standard deviation measures variability by calculating the average squared distance of all data observations from the mean of the dataset. When this metric is a lower number (relative to the mean of the dataset), it indicates that most of the data values are clustered closely together, whereas a higher value indicates a higher level of variation and spread. Note, though, that a low or high standard deviation value depends on the dataset at hand. If the dataset has a range of 200 and a standard

---

[22] Spread or dispersion refers to the variability and spread of the data values.
[23] Sarah Boslaugh, "Statistics in a Nutshell," *O'Reilly Media*, Second Edition, 2012.

deviation of 25, then the variation is high, but if the range of that dataset was 2,000, then the variation would be considered low.

Dataset 1	Dataset 2	Dataset 3
4	2	1
5	2	2
3	10	1
12	10	20
Mean: 6	Mean: 6	Mean: 6
Range: 9	Range: 8	Range: 19
Standard Deviation: 4.082	Standard Deviation: 4.618	Standard Deviation: 9.345

Table 8: The mean, range, and standard deviation of three sample datasets

To better understand standard deviation, let's explore the three datasets in Table 8. As we can see, each dataset has a mean of 6. In addition, Dataset 1 and Dataset 2 have a similar range. Yet, there is more variability between these three datasets than the **mean** and **range** suggest. We can expand our analysis using standard deviation.

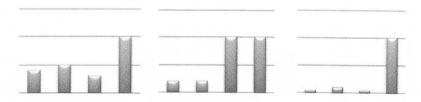

Figure 6: Dataset 1, 2 & 3 plotted on bar graphs (scaled to size)

Dataset 1 has the lowest standard deviation score of 4.082. This indicates that Dataset 1 has the least variance between its four data points. Remember, the lower the standard deviation, the less variation in the data. Dataset 3 has the highest variance as indicated by its comparatively high standard deviation score of 9.345.

How to Calculate Standard Deviation

Standard deviation is calculated as follows:

$$\sqrt{\frac{1}{N-1} \sum_{i=1}^{N} (x_i - \bar{x})^2}$$

Where:

N = number of items contained in the dataset

Σ = sum (adding together), also known as the "sigma"

i = the range over which the operation should be performed, which in this equation is sum (Σ) of all the values of x from **1** to **N**.

x_i = an individual observation from the dataset[24]

x = the mean

Let's apply this formula using the following dataset as sample data.

Sample Dataset
4
7
5
12
Mean = 7

---

[24] · The symbol i designates the position in the dataset, so x_1 is the first value and x_2 is the second value in the dataset.

Table 9: Sample dataset

n = **4** (the dataset has four items)

x_i = **4, 7, 5, 12** (x_1, x_2, x_3, and x_4)

x = **7** (4 + 7 + 5 + 12 = 28/4)

Next, input this information into the standard deviation formula, as shown here.

$$\sqrt{\frac{1}{4-1}\sum_{i=1}^{4}(x_i-7)^2}$$

We then input each **observation x_i** (**"4," "7," "5," "12"**) from the dataset into the equation to find the sum (Σ). The "$i=1$" below the sigma and N (4) notes that we need to add each observation from the dataset separately into the equation and then sum (Σ) the total. This is demonstrated below.

(4 − 7)² + **(7** − 7)² + **(5** − 7)² + **(12** − 7)²

(-3)² + (0)² + (-2)² + (5)²

9 + 0 + 4 + 25 = 38

$$\sqrt{\frac{1}{3}(38)}$$

Next, multiply 38 by the fraction of one over three.

$$\sqrt{12.666}$$

Apply the square root function. You may wish to use an online tool such as Calculate Soup to find the square root (https://www.calculatorsoup.com/calculators/algebra/squareroots.php).

= 3.588

Round the final value to three decimal spaces (or as many decimal places as you choose) to produce a final standard deviation score. The standard deviation for this sample dataset is 3.558.

The Empirical Rule

A useful visual technique for interpreting data variance is to plot the dataset's distribution values using a histogram. A normal distribution, where data is distributed evenly, produces a bell curve (called a "bell curve" because it looks like a bell in shape) as shown in Figure 7. A normal distribution with a mean of 0 and a standard deviation of 1 is also known as a *standard normal distribution*. Normal distribution can be transformed to a standard normal distribution by converting the original values to standardized scores.

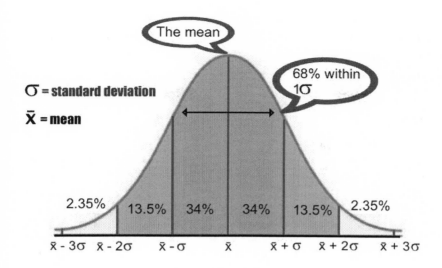

Figure 7: A symmetrical bell curve of a standard normal model

Besides the bell-shaped curve, normal distribution has a number of other recognizable features. For one, the highest point of the dataset occurs at the mean (x). Second, the curve is symmetrical around an imaginary line that lies at the mean. At its outermost ends, the curves approach but never quite touch or cross the horizontal axis. Lastly, the location at which the curves transition from upward to downward cupping (known as inflection points) occur one standard deviation above and below the mean.

The symmetrical shape of normal distribution is a reasonable description of how variables often diverge in the real world. From body height to IQ tests, variable values generally gravitate towards a symmetrical shape around the mean as more cases are added.[25] This phenomenon has been quantified under what's called the *Empirical Rule*,[26] which describes normal distribution as follows:

---

[25] This phenomenon is similar to how parked cars fill in evenly on both sides of the entrance of a large department store, with a lower frequency of cars parked on the far left and far right of the car lot.

[26] All normal distributions, irrespective of their mean and standard deviation,

Approximately 68% of values fall within one standard deviation of the mean.

Approximately 95% of values fall within two standard deviations of the mean.

Approximately 99.7% of values fall within three standard deviations of the mean.

Known also as the *68 95 99.7 Rule* or the *Three Sigma Rule*, the Empirical Rule was coined by the French mathematician and consultant to gamblers Abraham de Moivre. Following an empirical experiment flipping a two-sided coin, de Moivre discovered that an increase in events (coin flips) gradually leads to a symmetrical curve of binomial distribution.[27] (Binomial distribution is used to describe a statistical scenario when only one of two mutually exclusive outcomes of a trial is possible, i.e., a head or a tail, true or false.)

Notice, for example, the gradual change in the shape of distribution of flipping a head using a two-sided coin in the following three scenarios (two coin flips, three coin flips, four coin flips).

---

generally follow the 68, 95, 99.7 rule.

[27] · Binomial distribution is commonly found in many real-life events such as student admission (accepted, not accepted), A/B testing, and computer database security (secure, security breach).

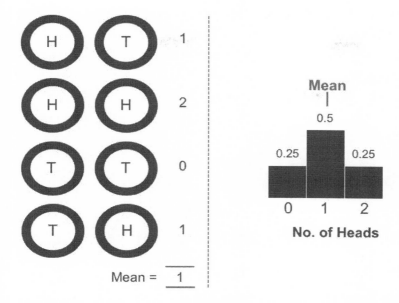

Figure 8: Total possible outcomes of flipping a head using two standard coins

In Figure 8, the diagram logs all possible outcomes of flipping two coins. The possibility of flipping at least one head is 50% and flipping either two heads or two tails is 25% respectively.

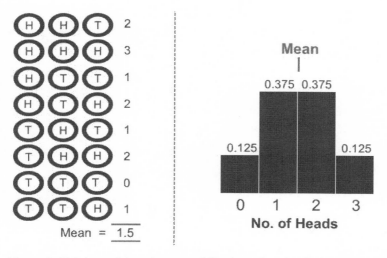

Figure 9: Total possible outcomes of flipping a head with three standard coins

After a third coin is added, the histogram expands to four possible outcomes, and 75% of the total dataset is in the center.

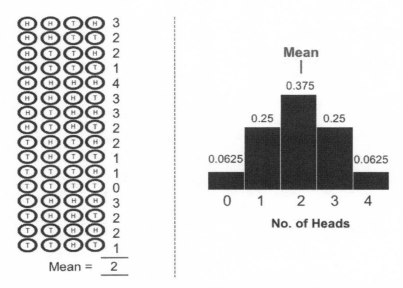

Figure 10: Total possible outcomes of flipping a head with four standard coins

With four coins included in the experiment, the histogram expands to five possible outcomes and the probability of most outcomes is now lower. With more data, the histogram contorts into a symmetrical bell-shape. As more data is collected, more observations settle in the middle of the bell curve, and a smaller proportion of observations land on the left and right tails of the curve. The histogram eventually produces approximately 68% of values within one standard deviation of the mean.

Using the histogram, we can pinpoint the probability of a given outcome such as two heads (37.5% in Figure 10) and whether that outcome is common or uncommon compared to other results—a potentially useful piece of information for gamblers and other prediction scenarios. It's also interesting to note that the mean, median, and mode all occur at the same point on the curve as this location is both the symmetrical center and the most common point. However, not all frequency curves produce a normal distribution as we shall explore next.

MEASURES OF POSITION

A large part of inferential statistics is determining the probability of a result—especially when that result appears extreme and beyond the occurrence of random chance.

We know on a normal curve there's a decreasing likelihood of replicating a result the further that observed data point is from the mean. We can also assess whether that data point is approximately one (68%), two (95%) or three standard deviations (99.7%) from the mean. This, however, doesn't tell us the probability of replicating the result. While it helps to map the overall variance of the dataset's data points, the standard deviation doesn't explain the actual variance of an individual data point from the mean.

Depending on the size of the dataset, there are two methods to identify the probability of replicating a result. The first method is to use the *Z-Score*, which finds the distance from the sample's mean to an individual data point expressed in units of standard deviation.[28]

---

[28] Transforming each of the recorded values in a dataset to a Z-Score converts a normally distributed population into a standard normal distribution with a mean of 0.

$$Z = \frac{(x_i - \bar{x})}{S}$$

Where:

x_i = an individual observation from the dataset

x = the mean of the sample

s = standard deviation of the population

Sample Dataset	Mean	Standard Deviation
4,7,3,5,2,6,7,4,2,6,8,12,7,3,5,5,3,1,6,3,6,4,3,4,8,9,4,5,2,6	5	2.366

Table 10: Sample dataset

Let's apply this formula using the first data observation (**4**) from the sample dataset. The dataset has a mean value of 5 and a standard deviation of **2.366**. Let's input these values into the formula.

$$Z = \frac{4 - 5}{2.366}$$

Z = (4 – 5) / 2.366

Z = -1 / 2.366

Z-Score = -0.42 (rounded to two decimal places)

The Z-Score for the first observation is -0.42, which means the data point is positioned 0.42 standard deviations from the mean in the negative direction, which is logical given this data point is lower than the mean (5).

Next, let's apply the same formula using a different observation from the sample dataset. This time, let's try "12." The values for the mean and standard deviation remain the same as with the previous example.

$$Z = \frac{12 - 5}{2.366}$$

Z = (12 – 5) / 2.366

Z = 7 / 2.366

Z-Score = 2.96 (rounded to two decimal places)

This time the Z-Score is 2.96, which means the data point is located 2.96 standard deviations from the mean in the positive direction. This data point could also be considered an anomaly as it is close to three deviations from the mean and different from other data points.

Outliers and Anomalies

After calculating the standard deviation of each observation from the mean, we can better determine whether an individual observation qualifies as an outlier or an anomaly, which refers to data points that lie an abnormal distance from other data points.

It's important to differentiate between anomalies and outliers. An anomaly is a rare event that is abnormal and perhaps should not have occurred. In the case of a normal distribution, we can instantly recognize whether a data point qualifies as an anomaly. If the Z-Score falls three positive or negative deviations from the mean of the dataset, it's fair to say that value is an anomaly because it falls beyond 99.7% of the other data points on a normal distribution curve. In some cases, it's viewed as a negative exception, such as fraudulent behavior or an environmental crisis.

Outliers are closely linked but represent a slightly larger grouping of data points than isolated anomalies. While there isn't a unified agreement on how to define outliers, statisticians generally view data points that diverge from primary data patterns as outliers because they record unusual scores on at least one variable and are more plentiful than anomalies.

Both outliers and anomalies can distort your predictions—even with a minimal number of cases—but they can also be viewed in a positive light. Anomalies help to identify data entry errors and are commonly used in fraud detection to identify illegal activities. In other cases, you may wish to study why outliers are different as popularized by Malcolm Gladwell's best-selling book *Outliers: The Story of Success*.

T-Score

We've covered normal distribution and the Empirical Rule (68, 95, 99.7), and we've applied the Z-Score to examine how many standard deviations a given data point is from the mean. The Z-Score enables us to identify whether a data point is, say, an anomaly or close to the mean but what wasn't mentioned was that the Z-Score applies to a normally distributed sample with a known standard deviation of the population. However, sometimes the mean isn't normally distributed or the standard deviation of the population is unknown or not reliable, which could be due to insufficient sampling.

Size	Sample Dataset	Mean	Standard Deviation
13	4,7,3,5,2,6,7,4,2,6,8,12,7	5.6	2.6
30	4,7,3,5,2,6,7,4,2,6,8,12,7,3,5,5,3,1,6, **3,6,4,3,4,8,9,4,5,2,6**	5	2.366

Table 11: The standard deviation of small datasets is susceptible to change as more observations are added

How can we know that the mean and standard deviation of a small sample will reflect the true population as more observations are added? The answer is we *don't*. Thus, rather than calculate the Z-Score using Z-distribution, we need to find what's called the *T-Score* using *T-distribution*, which was developed in the early 20th Century by the Irish statistician W. S. Gosset.

Not wishing to jeopardize his job at the local brewery, Gosset published under the pen name "Student," which explains why T-distribution is sometimes called "Student's T-distribution."

The Z-Score and T-Score perform the same primary function (measure distribution), but like two screwdrivers with slightly different heads, they're used with different sizes of sample data. The Z-Score, as we've covered, measures the deviation of an individual data point from the mean for datasets with 30 or more observations based on Z-distribution (standard normal distribution).

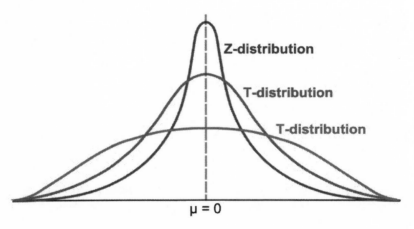

Figure 11: Z-distribution and T-distribution

Unlike Z-distribution, T-distribution is not one fixed bell curve. Its distribution curve changes in accordance with the size of the sample. In cases where the sample size is small, e.g. 10, the curve appears relatively flat with a high proportion of data points in the curve's tails. As the sample size increases, the distribution curve approaches the standard normal curve (Z-distribution) with more data points closer to the mean at the center of the curve.

A standard normal curve, as you'll recall, is defined by the 68 95 99.7 rule, which sets approximate confidence levels for one, two, and three standard deviations from a mean of 0. Based on this rule, 95% of data points will fall 1.96 standard deviations from the mean.[29] Thus, if the sample's mean is 100 and we randomly select an observation from the sample, the probability of that data point falling within 1.96 standard deviations of 100 is 0.95 or 95%. To find the exact variation of that data point from the mean we can use the Z-Score. In the case of smaller datasets—that don't follow a normal curve—we instead need to use the T-Score.

$$T = \frac{\bar{X} - \mu_0}{\frac{s}{\sqrt{n}}}$$

Where:

x = the mean of the sample

μ_0 = the mean of the sample population

s = standard deviation of the sample

n = sample size

The formula is similar to that of the Z-Score, except the standard deviation is divided by the sample size. Also, the standard deviation relates to the sample in question, which may or may not reflect the standard deviation of the population (which contains more data points).

---

[29] The Z-Score is found by referencing an online calculator or Z-distribution table, available in Chapter 9, Table 18 as an excerpt example.

Sample Dataset	Mean	Standard Deviation
4,7,3,5,2,6,7,4,2,6,8,12,7,3,5,5,3,1,6	5.05	2.592

Table 12: A sample dataset

Let's use this formula to find the T-Score of the first data point from the sample dataset shown in Table 12. Note that the dataset has a mean of 5.05 and a standard deviation of 2.592. Let's input these values into the T-Score formula.

$$T = \frac{4 - 5.05}{\frac{2.592}{\sqrt{19}}}$$

T = (4 – 5.05) / (2.592 / √19)

T = -1.05 / (2.592 / 4.359)

T = -1.05 / 0.595

T-Score = -1.765

According to the T-Score, the sample data point is positioned -1.765 standard deviations from the sample's mean. As T-distribution does not follow the 68, 95, 99.7 rule of distribution, we need to use a T-table to find the probability of this result.

Degrees of Freedom	Total area in one tail (a)					
	0.001	0.005	0.010	0.025	0.050	0.100
1	318.3088	63.65674	31.82052	12.70620	6.313752	3.077684
2	22.32712	9.924843	6.964557	4.302653	2.919986	1.885618
3	10.21453	5.840909	4.540703	3.182446	2.353363	1.637744
4	7.173182	4.604095	3.746947	2.776445	2.131847	1.533206
5	5.893430	4.032143	3.364930	2.570582	2.015048	1.475884
6	5.207626	3.707428	3.142668	2.446912	1.943180	1.439756
7	4.785290	3.499483	2.997952	2.364624	1.894579	1.414924
8	4.500791	3.355387	2.896459	2.306004	1.859548	1.396815
9	4.296806	3.249836	2.821438	2.262157	1.833113	1.383029

Table 13: T-distribution table

Listed from left to right in the T-distribution table head are the alpha values. On the vertical axis, we can see the degrees of freedom (df) listed from 1 to 9. Degrees of freedom (df) is the sample size minus 1. A sample size of 10, for example, has 9 degrees of freedom.

Sample Size	Degrees of Freedom
10	9
15	14
100	99
1000	999

Table 14: Examples of degrees of freedom

Let's now look up our T-Score of negative 1.765 in the T-distribution table. In this table, we can't actually find a T-Score of 1.765 with 4 degrees of freedom (5 − 1). However, we can see that the T-Score falls between 1.53 and 2.13, which accounts for

90% (alpha of 0.100) to 95% (alpha of 0.050) of the total distribution. This also means that our result (4) is not particularly unusual in relation to the mean of our sample (5.05).

Degrees of Freedom	Total area in one tail (a)					
	0.001	0.005	0.010	0.025	0.050	0.100
1	318.3088	63.65674	31.82052	12.70620	6.313752	3.077684
2	22.32712	9.924843	6.964557	4.302653	2.919986	1.885618
3	10.21453	5.840909	4.540703	3.182446	2.353363	1.637744
4	7.173182	4.604095	3.746947	2.776445	2.131847	1.533206

Table 15: T-distribution table excerpt

Both T-distribution and Z-distribution play an important role in hypothesis testing and we will revisit these measures of variance in later chapters. In the next chapter, we'll look at the basic mechanics of hypothesis testing.

8

THE HYPOTHESIS TEST

Working for a tech company with roots in e-commerce and a large stake in AI, we often used hypothesis testing on user behavior.

"Should we place the 'Buy Now' button on the left-hand side or the right-hand side of the website banner?"

"Should the text for that button be 'Get Started' or 'Free Trial'?"

These questions are frequently asked by promotion managers and examined by the user experience team. Colors are another source of debate and regular hypothesis testing.

Let's suppose you join our team, and after settling into your cubicle you suggest that **green** would be a more effective color than **blue** as a call-to-action (CTA) button. Call-to-action buttons include the "Contact Us," "Buy Now," and "Free Trial" buttons on a website which lead to important user actions on the site. From the company's perspective, conversions from these buttons generate customer leads, sales, revenue, and customer retention.

Figure 12: Examples of website call-to-action buttons

However, as a new member of the team, it's difficult to validate your hypothesis to a group of established decision-makers without solid evidence. Your best bet is to find and use data to test your hypothesis.

The first step is to set up your hypothesis statement. This takes the form of two hypotheses: H_0 and H_1. H_0 is the null hypothesis, which declares that changing the call-to-action buttons from blue to green has no positive impact on user conversion. User conversion refers to the percentage of users who click the button. The alternative hypothesis, H_1, states the opposite and that changing from blue to green will have a positive effect.

One-Tailed vs Two-Tailed

There are two primary methods of hypothesis testing, and they are known as one-tailed and two-tailed tests. In a nutshell, a one-tailed test considers one direction of results (left or right) from the null hypothesis, whereas a two-tailed test considers both directions (left and right).

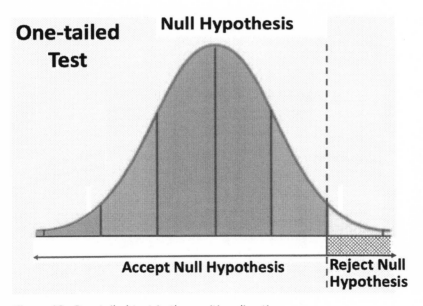

Figure 13: One-tailed test in the positive direction

Our experiment, which analyzes conversion rates for an e-commerce website's call-to-action button, qualifies as one-tailed testing. The objective of the test is to dispel/nullify the null hypothesis, which is that blue leads to more clicks. Thus, if the experiment's results for green return a conversion rate lower than or equal to the conversion rate set by blue, the null hypothesis stands to be true. If the experiment results return a positive conversion—greater than say 3 standard deviations—the null hypothesis is rejected. Hence, the rejection of the null hypothesis can only be triggered in one direction. Anything less than +3 standard deviations, in this example, holds that the null hypothesis is true and that blue is the most effective color for user conversion. Conversely, a two-tailed test considers both directions from the null hypothesis.

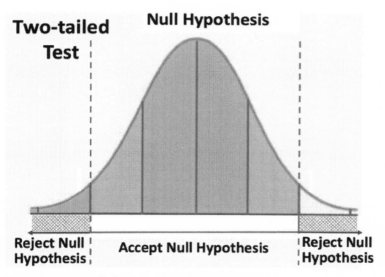

Figure 14: Two-tailed test in the left and right direction

To illustrate this second method of testing, let's imagine that our hypothesis test aims to challenge the current conversion rate of 30%. This means that on each page of the website, only 30% of users click on a blue button to complete a purchase or respond to another call-to-action, such as "Contact Sales" or "Register an Account."

However, rather than claim that an alternative color will raise conversions, our experiment's goal is to cast doubt and nullify the claimed accuracy of the current conversion rate. In other words, we disagree that 30% is the actual conversion rate of the blue button, and we think the real number is either significantly higher or lower than 30%.

The rationale behind this hypothesis is based on our knowledge that the website isn't available on certain browser types and conversion buttons are not rendering on the screen for some users. The fact that the conversion buttons are not displaying for all users could be deflating the overall conversion rate. In addition, there may be a problem with the analytics software the company is using to analyze web traffic, which isn't recording results for users of certain browser types, such as Safari.

Thus, the objective of the hypothesis test is not to challenge the null hypothesis in one particular direction but to consider both directions as evidence of an alternative hypothesis. For this test, the alternative hypothesis is accepted if the conversion rate is significantly higher or lower than the null hypothesis of 30%. In Figure 14 can see there are two rejection zones, known as the *critical areas*. Results that fall within either of the two critical areas trigger a rejection of the null hypothesis and thereby validate the alternative hypothesis.

Type I and Type II Errors

Before we proceed further, be wary that neither the null hypothesis or the alternative hypothesis can be unequivocally proven correct in hypothesis testing. Analyzing a sample extracted from a larger population is a subset of the data, and thus, any conclusions formed about the larger population based on analyzing the sample data are considered probabilistic rather than absolute. The sample data can only support that the null hypothesis or the alternative hypothesis is probable with a minimal error of interpretation. However, mistakes do occur, and there are two main types of errors that we need to manage and minimize.

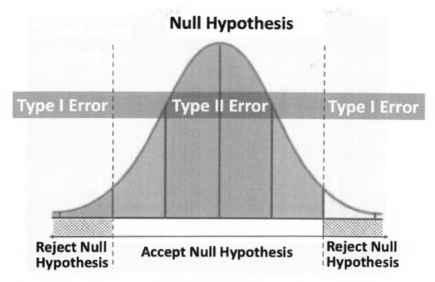

Null Hypothesis

Type I Error   Type II Error   Type I Error

Reject Null Hypothesis   Accept Null Hypothesis   Reject Null Hypothesis

Figure 15: Example of Type I and Type II Errors in hypothesis testing

The first mistake is a Type I Error, which is the rejection of a null hypothesis (H_0) that was true and should not have been rejected. This means that although the data appears to support that a relationship is responsible, the covariance (a measurement of how related the variance is between two variables) of the variables is occurring entirely by chance. Again, this does not prove that a relationship doesn't exist, merely that it's not the most likely cause. This is commonly referred to as a *false-positive*.

Conversely, a Type II Error is accepting a null hypothesis that should've been rejected because the covariance of variables was probably not due to chance. This is also known as a *false-negative*.

Let's use a pregnancy test as an example. In this scenario, the null hypothesis (H_0) holds that the woman is not pregnant. The null hypothesis is therefore **rejected** if the woman is pregnant (H_0 is false) and **accepted** if the woman is indeed not pregnant (H_0 is true).

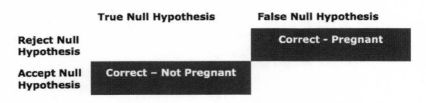

Table 16: Example of a true positive and true negative

However, in some cases, the test may not be 100% accurate and mistakes may occur. If the null hypothesis is **rejected** and the woman is not actually pregnant (H_0 is true), this leads to a Type I Error. If the null hypothesis is **accepted** and the woman is pregnant (H_0 is false), this leads to a Type II Error.

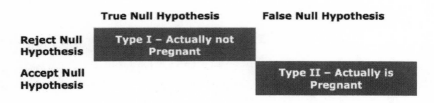

Table 17: Example of Type I and Type II Errors

Statistical Significance

Given that the sample data cannot be truly reliable and representative of the full population, there's the possibility of a sampling error or random chance affecting the experiment's results. To better understand this phenomenon, let's examine another practical example.

Let's imagine we want to study online gaming habits in Europe and our null hypothesis for the average time spent on mobile games among teenagers in Europe is 15 hours per week. For this experiment, we want to test the validity of the null hypothesis. Obviously, it would be extremely complicated and difficult to analyze the gaming habits of every teenager across many gaming platforms. We'll therefore use data sampling and inferential statistical analysis to run the experiment. Extracted from the full population of all teenagers in Europe, we'll randomly select 100 respondents and monitor their gaming behavior over a seven-day period.

After a week of observing their behavior, we receive the results. The mean time spent gaming by the sample is 22 hours per week. While some teenagers in the study reported no hours gaming and others played daily, the figure of 22 hours represents the average time spent across the entire sample. Based on the results of this experiment, it's natural to doubt the validity of the null hypothesis, which states that teenagers in Europe spend an average of 15 hours a week gaming—a mean value much lower than our sample mean.

However, if we were to repeat the experiment, with a new batch of 100 randomly selected teenagers, could we rely on the results staying the same? Alternatively, would the results edge closer to the null hypothesis of 15 hours?

Herein lies the trouble and complexity of hypothesis testing: not all samples randomly extracted from the population are preordained to reproduce the same result. It's natural for some samples to contain a higher number of outliers and anomalies than other samples, and naturally, results can vary. If we continued to extract random samples, we would likely see a range of results and the mean of each random sample is unlikely to be equal to the true mean of the full population.

Returning to our experiment, we now need to ask ourselves, can we dispel the null hypothesis based on the first sample result? Yes, the sample mean is different from the null hypothesis but *how* different? This leads us to statistical significance which outlines a threshold for rejecting the null hypothesis.

Statistical significance is often referred to as the *p-value* (probability value) and is expressed between 0 and 1. A p-value of 0.05, for example, expresses a 5% possibility of replicating a result if we take another sample.

In hypothesis testing, the p-value is compared to a pre-fixed value called the alpha. If the p-value returns as equal or less than alpha, then the result is statistically significant and we can reject the null hypothesis. If the p-value is greater than alpha, the result is not statistically significant and we cannot reject the null hypothesis. Alpha thus sets a fixed threshold for how extreme the results must be before rejecting the null hypothesis.

It's important to underline that the alpha should be defined before the experiment and not after the results have been obtained. For two-tailed tests, the alpha is divided by two. Thus, if the alpha is 0.05 (5%), then the critical areas of the curve each represent 0.025 (2.5%). Also, although hypothesis tests usually adopt an alpha of between 0.01 (1%) and 0.1 (10%), there is no predefined or optimal alpha for all hypothesis tests.

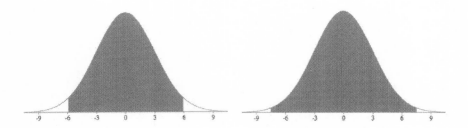

Figure 16: Confidence of 95% on the left, and confidence of 99% on the right

Given that alpha is equal to the probability of a Type I Error (incorrect rejection of the null hypothesis), there's a natural tendency to set alpha to a low value such as 0.01 to avoid a Type I Error. That way, the critical area is smaller, and there's less chance of incorrectly rejecting the null hypothesis. While this sounds good in theory, reducing the risk of Type I Error increases the risk of a Type II Error (incorrectly accepting the null hypothesis). This two-sided scenario represents an inherent trade-off in hypothesis testing.

As a way to mitigate the problem, most industries have found that 0.05 (5%) is the ideal alpha for hypothesis testing.[30] While a lower alpha of $p < 0.01$ is sometimes used as an alternative, there is little evidence supporting a high alpha such as $p < 0.10$.[31]

According to Sarah Boslaugh, " there's nothing magical about the 0.05 level." In her book *Statistics in a Nutshell*, Boslaugh explains that the 0.05 alpha was adopted in the early 20th Century when statistics were computed by hand and compared to published tables with fixed significance levels.[32] For now, though, 0.05 remains the default standard for hypothesis testing.

Confidence Levels

An experiment's alpha also determines "confidence," a concept introduced earlier. Expressed between 0% and 100%, confidence is a statistical measure of prediction confidence regarding whether the sample result of the experiment is true of the full population.

Confidence is calculated as $(1 - a)$. If the alpha is 0.05, for example, then the confidence level of the experiment is 0.95 (95%).

$1.0 - a$ = confidence level

$1.0 - 0.05 = 0.95$

We now know how to calculate confidence levels using alpha and to reject the null hypothesis if the sample is statistically significant. The next step is to define the critical areas set by alpha. An alpha of 0.05 holds that we reject the null hypothesis when the results are in a 5% zone, but this doesn't tell us where to plant the null hypothesis rejection zone(s). Instead, critical

---

[30] J.H Zar, "Biostatistical Analysis," *Prentice-Hall International*, New Jersey, 1984.
[31] Sarah Boslaugh, "Statistics in a Nutshell," *O'Reilly Media*, Second Edition, 2012.
[32] Sarah Boslaugh, "Statistics in a Nutshell," *O'Reilly Media*, Second Edition, 2012.

areas can be found on the curve by finding the confidence intervals.

Confidence intervals define the confidence bounds of the curve, which is one or two intervals depending on the type of the experiment. In the case of a two-tailed test, two confidence intervals define two critical areas outside the upper and lower confidence limits, whereas, for a one-tailed test, a single confidence interval defines the left/right-hand side critical area.

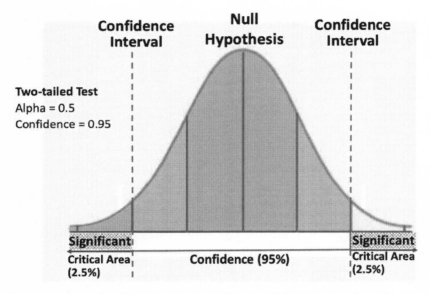

Figure 17: Example of a two-tail test with two confidence intervals and two critical areas

Z-distribution Confidence Intervals

While there are numerous methods to calculate the confidence intervals depending on the type of test (left one-tailed, right one-tailed, two-tailed) and the sample data (n) available, let's look at a case of normal distribution where there is sufficient sample data (n>30) for a two-tailed test using the following formula:

$$\bar{x} \pm Z\left(\frac{\sigma}{\sqrt{n}}\right)$$

Where:

x = mean of the sample

± = plus-minus sign

Z = Z-distribution critical value

σ = standard deviation of the sample dataset

n = sample size

The Z-Statistic is used to find the distance between the null hypothesis and the sample mean. In hypothesis testing, the experiment's Z-Statistic is compared with the expected statistic (called the *critical value*) for a given confidence level.

To experiment with calculating confidence intervals, let's first use the sample data from our experiment examining teenage gaming habits in Europe.

Example

A group of 100 teenagers had a mean of 22 hours gaming with a standard deviation of 5.7. Using a two-tailed test, let's find the confidence intervals for 95% confidence with an alpha of 0.05.

Test Type: Two-tailed.

x = 22 (hours)

Z = 1.96, Z is found using a Z-distribution table

σ = 5.7 (standard deviation)

n = 100

$$\bar{x} \pm Z\left(\frac{\sigma}{\sqrt{n}}\right)$$

22 ± 1.96 x (5.7 / √100)

22 ± 1.96 x (5.7 / 10)

22 ± 1.96 x (0.57)

22 ± 1.1172

22 – 1.1172 = 20.8828 (lower interval)

22 + 1.1172 = 23.1172 (upper interval)

Based on this upper and lower confidence intervals, we're saying that we're 95% certain that our sample data will fall somewhere between 20.8828 and 23.1172 hours.

T-distribution Confidence Intervals

Let's now repeat this exercise using a smaller sample size and T-distribution to set our confidence intervals.

Example

A group of 9 teenagers recorded a mean of 22 hours gaming with a standard deviation of 5 hours. Let's find the confidence interval for 95% confidence with an alpha of 0.05.

Test Type: Two-tailed

x = 22 (hours)

T = 2.306 (T critical score with an alpha of 0.025 in each tail and 8 df)

σ = 5 (standard deviation)

n = 9

$$\bar{x} \pm T \left(\frac{\sigma}{\sqrt{n}} \right)$$

22 ± 2.306 x (5 / √9)

22 ± 2.306 x (5 / 3)

22 ± 2.306 x (1.666)

22 ± 3.842

22 − 3.842 = 18.158 (lower interval)
22 + 3.842 = 25.842 (upper interval)

Summing Up

Although it can seem confusing and there are many terms to learn and understand, the overall objective of hypothesis testing is to prove that the outcome of the sample data is representative of the full population and not occurring by chance caused by randomness in the sample data. Understanding statistical significance and setting a reasonable alpha for defining your confidence intervals/critical areas are both important steps for mitigating this potential problem.

In summary, hypothesis testing can also be broken down into these four steps:

Step 1: Identify the null hypothesis (what you believe to be the status quo and wish to nullify) and the type of test (i.e. one-tailed or two-tailed).

Step 2: State your experiment's alpha (statistical significance and the probability of a Type I Error) and set the confidence interval(s).

Step 3: Collect sample data and conduct a hypothesis test.

Step 4: Compare the test result to the critical value (expected result) and decide if you should support or reject the null hypothesis.

In the next chapter, we'll begin introducing specific methods of hypothesis testing and evaluating the critical value as part of Step 3 and Step 4 respectively.

9

THE Z-TEST

Using the alpha/statistical significance, confidence intervals, and critical areas discussed in the last chapter, let's explore hypothesis testing using the Z-Test.

The Z-Test is based on Z-distribution as introduced in Chapter 7. However, rather than using the **Z-Score** to measure the distance between a data point and the sample's mean, in hypothesis testing, we use what's called the **Z-Statistic** to find the distance between a sample mean and the null hypothesis.

Like the Z-Score, the Z-Statistic is expressed numerically, and the higher the statistic, the higher the discrepancy there is between the sample data and the null hypothesis. Conversely, a Z-Statistic of close to 0 means the sample mean matches the null hypothesis—confirming the null hypothesis.

Every Z-Statistic is also pegged to a p-value, which is the probability of that result occurring by chance. P-values range from 0% to 100% and are usually expressed in decimal form, i.e. 5% = 0.05. A low p-value, such as 0.05, indicates that the sample mean is unlikely to have occurred by chance. As discussed, in many experiments, a p-value of 0.05 is sufficient to reject the null hypothesis.

To find the p-value for a Z-statistic, we need to refer to a Z-distribution table as shown in Table 18.

Confidence	Z
90%	1.645
95%	1.96
98%	2.326
99%	2.576

Table 18: Excerpt of a Z-distribution table

There are numerous tests to find the Z-Statistic and in this chapter we will discuss two-sample and one-sample testing.

Two-sample Z-Test

A two-sample Z-Test compares the difference between the means of two independent samples with a known standard deviation. This test is also based on the assumption that the data is normally distributed with a minimum of 30 observations.

$$Z = \frac{(\bar{X}_1 - \bar{X}_2) - (\mu_1 - \mu_2)}{\sqrt{\frac{s_1^2}{n_1} + \frac{s_2^2}{n_2}}}$$

Where:

x_1 = mean of first sample

x_2 = mean of second sample

$\mu_1 - \mu_2$ = null hypothesis which is usually set to 0 (no difference between the two samples)

s_1 = standard deviation of first sample

s_2 = standard deviation of second sample

n_1 = size of first sample

n_2 = size of second sample

Example

A new smartphone company named **Company A** claims their phone battery outperforms its main competitor called **Company B**. A random sample of 60 users is enlisted to test this claim. The mean battery life of Company A among a sample of 30 users was 21 hours, with a standard deviation of 3. Company B had a mean of 19 hours among 30 users, with a standard deviation of 2.

Is their evidence to support company A's claim? Let's test the difference between the two samples using a right-tailed test with an alpha of 0.05 (95% confidence).

H_0: $\mu_1 = \mu_2$

H_1: $\mu_1 > \mu_2$

$$Z = \frac{(\bar{X}_1 - \bar{X}_2) - (\mu_1 - \mu_2)}{\sqrt{\frac{s_1^2}{n_1} + \frac{s_2^2}{n_2}}} = \frac{(21 - 19) - (0 - 0)}{\sqrt{\frac{9}{30} + \frac{4}{30}}}$$

$Z = ((21 - 19) - (0 - 0)) / \sqrt{(3^2/30 + 2^2/30)}$

$Z = 2 / \sqrt{(9/30 + 4/30)}$

$Z = 2 / \sqrt{0.433}$

$Z = 2 / 0.658$

Z-Statistic $= 3.039$

Let's now compare the Z-Statistic with the expected critical score of 95% confidence. According to the Z-distribution table, 0.95% of data points should fall within 1.645 standard deviations of the null hypothesis for a one-tailed test. (Note that the critical score of 1.645 for a one-tailed test is the same critical score as that of a two tailed-test with 90% confidence).[33] Thus, with our Z-Statistic of 3.039, we can conclude that the samples are in fact different and we can reject the null hypothesis of $\mu_1 = \mu_2$.

---

[33] As we are only analyzing one tail, we need to double the range of the critical area by doubling the alpha from 0.05 (95% confidence) to 0.1 (90% confidence).

One-sample Z-Test

Let's now look at a one-sample Z-Test to test the validity of a stated mean. This test is also based on the assumption that the data is normally distributed with a known standard deviation and a minimal sample size of 30.

$$Z = \frac{\bar{x} - \mu_0}{\frac{s}{\sqrt{n}}}$$

Where:

x = mean of the sample

μ_0 = hypothesized mean (null hypothesis)

s = standard deviation

n = size of sample

Example

The same smartphone company claims their new phone battery outperforms the 20 hour average of other smartphones. A random sample of 30 users is asked to test the claim using the new smartphone. The mean battery time among the sample users was 21 hours with a standard deviation of 3 hours. Is there sufficient evidence to reject the company's marketing claim?

Let's test the difference of the sample mean and the known mean using a right-tailed test with an alpha of 0.05 (95% confidence).

H_0: $\mu_1 = 20$

H_1: $\mu_1 > 20$

$$Z = \frac{\bar{x} - \mu_0}{\frac{s}{\sqrt{n}}} = \frac{21 - 20}{\frac{3}{\sqrt{30}}}$$

$Z = (21 - 20) / (3/\sqrt{30})$

$Z = 1 / (3/5.477)$

$Z = 1 / 0.547$

Z-Statistic $= 1.828$

Based on an expected critical value of 1.645 for a right-tailed test with 95% confidence, we can conclude that a Z-Statistic of 1.828 standard deviations from the null hypothesis is statistically significant. The samples are therefore sufficiently different to warrant rejecting the null hypothesis of $\mu_1 = 20$.

10

THE T-TEST

Similar to the Z-Test, a T-Test analyzes the distance between a sample mean and the null hypothesis but is based on T-distribution (using a smaller sample size) and uses the standard deviation of the sample rather than that of the population.

There are three main types of T-Tests:

- An **independent samples T-Test** (two-sample T-Test) for comparing means from two different groups, such as two different companies or two different athletes. This is the most commonly used type of T-Test.

- A **dependent sample T-Test** (paired T-test) for comparing means from the same group at two different intervals, i.e. measuring a company's performance in 2017 against 2018.

- A **one-sample T-Test** for testing the sample mean of a single group against a known or hypothesized mean.

The output of a T-Test is called the T-Statistic, which quantifies the difference between the sample mean and the null hypothesis. As the T-Statistic increases in the positive or negative direction, the gap in results between the sample data and null hypothesis expands. However, rather than refer to a Z-distribution table (as we did in the previous chapter) to find the expected critical value, we instead use a T-distribution table.

Degrees of Freedom	Total area in one tail (a)					
	0.0005	0.001	0.005	0.010	0.025	0.050
1	636.6192	318.3088	63.65674	31.82052	12.70620	6.313752
2	31.59905	22.32712	9.924843	6.964557	4.302653	2.919986
3	12.92398	10.21453	5.840909	4.540703	3.182446	2.353363
4	8.610302	7.173182	4.604095	3.746947	2.776445	2.131847
5	6.868827	5.893430	4.032143	3.364930	2.570582	2.015048
6	5.958816	5.207626	3.707428	3.142668	2.446912	1.943180
7	5.407883	4.785290	3.499483	2.997952	2.364624	1.894579
8	5.041305	4.500791	3.355387	2.896459	2.306004	1.859548
9	4.780913	4.296806	3.249836	2.821438	2.262157	1.833113

Table 19: Excerpt of a T-distribution table

If we have a one-tailed test with an alpha of 0.05 and sample size of 10 (df 9), we can expect 95% of samples to fall within 1.83 standard deviations of the null hypothesis. If our sample mean returns a T-Statistic greater than the critical score of 1.83, we can conclude the results of the sample are statistically significant and unlikely to have occurred by chance—allowing us to reject the null hypothesis.

In the case of a two-tail test, the alpha is divided by 2 (a/2), which means the alpha is 0.025 (0.05/2) for each tail. Thus, for a two-tailed test, the two critical areas would each account for 2.5% of the distribution based on 95% confidence with confidence intervals of -2.262 and +2.262 from the null hypothesis.

Next are the formulas and example scenarios for T-Testing.

1) Independent Samples T-Test

An independent samples T-Test compares means from two different groups.

$$T = \frac{(\bar{X}_1 - \bar{X}_2)}{\sqrt{\frac{s_p^2}{n_1} + \frac{s_p^2}{n_2}}}$$

Where:

X_1 = mean of sample 1

X_2 = mean of sample 2

n_1 = size of sample 1 (number of observations)

n_2 = size of sample 2 (number of observations)

s_p^2 = pooled standard deviation

Pooled standard deviation (s_p^2) is calculated separately as follows:

$$s_p^2 = \frac{(n_1 - 1)s_1^2 + (n_2 - 1)s_2^2}{n_1 + n_2 - 2}$$

Where:

n_1 = size of sample 1

n_2 = size of sample 2

s_1^2 = standard deviation of sample size 1 squared

s_2^2 = standard deviation of sample size 2 squared

Example

An e-commerce store decides to compare customer spending between the desktop version of their website and the mobile site.

Their results found that 25 desktop customers spent an average of $70 with a standard deviation of $15. Among mobile users, 20 customers spent $74 on average with a standard deviation of $25.

Let's test the difference of the sample mean and the known mean using a two-tail test with an alpha of 0.05 (95% confidence).

$H_0: \mu_1 = \mu_2$

$H_1: \mu_1 \neq \mu_2$

Let's first solve the pooled standard deviation:

$$s^2_p = \frac{(n_1 - 1)s_1^2 + (n_2 - 1)s_2^2}{n_1 + n_2 - 2} = \frac{(25 - 1)15^2 + (20 - 1)25^2}{20 + 25 - 2}$$

$((25 - 1)15^2 + (20 - 1)25^2) / (20 + 25 - 2)$

$((24)225 + (19)625) / 43$

$(5400 + 11875) / 43$

$17275 / 43 = 401.74$

$s^2_p = 401.74$

Solve the full equation:

$$T = \frac{(\bar{x}_1 - \bar{x}_2)}{\sqrt{\dfrac{s^2_p}{n_1} + \dfrac{s^2_p}{n_2}}} = \frac{(70 - 74)}{\sqrt{\dfrac{401.74}{25} + \dfrac{401.74}{20}}}$$

$T = (70 - 74) / \sqrt{(401.74/25 + 401.74/20)}$

$T = -4 / \sqrt{(16.0696 + 20.087)}$

$T = -4 / \sqrt{35.1566}$

$T = -4 / 5.929$

T-Statistic = -0.675

Degrees of Freedom	Total area in one tail (a)					
	0.0005	0.001	0.005	0.010	0.025	0.050
19	3.883	3.579	2.861	2.539	2.093	1.729
20	3.850	3.552	2.845	2.528	2.086	1.725

Table 20: Excerpt of a T-distribution table

Let's now compare the T-Statistic with the expected critical value based on 95% confidence. The degrees of freedom is assigned according to the smaller of the two samples, which in this case is 20. The intersection for 19 degrees of freedom with an alpha of 0.025 in each tail is 2.093. Thus, based on our T-Statistic of -0.675, we can conclude that the samples are not significantly different and therein accept the null hypothesis of $\mu_1 = \mu_2$.

2) Dependent Sample T-Test

A dependent sample T-Test is used for comparing means from the same group at two different intervals.

$$T = \frac{\bar{X}_D - \mu_0}{\frac{S_D}{\sqrt{n}}}$$

Where:

X_D = the mean of the difference between the two samples
μ_0 = the hypothesized mean (null hypothesis)
S_D = standard deviation of the difference between two samples
n = size of sample

S_D is calculated separately as follows:

$$S_D = \sqrt{\frac{\Sigma x^2 - \frac{(\Sigma x)^2}{n}}{n - 1}}$$

Where:

$\Sigma_x{}^2$ = the sum of each squared difference

Σx = the sum of each difference

n = size of sample

Example

A mobile fitness company wants to test whether their mobile app increases the number of user workouts each month. They have ten new users trial the app in January and compare the data with the same participant's workout frequency from the previous month. Using a two-tail test with an alpha of 0.05, we state our hypotheses as:

H_0: $\mu_1 = \mu_2$

H_1: $\mu_1 \neq \mu_2$

	Workouts December	Workouts January	Difference (x)	Difference Squared (x^2)
User 1	7	8	1	1
User 2	1	9	8	64
User 3	4	4	0	0
User 4	5	7	2	4
User 5	3	4	1	1
Total (Σ)			12	70

Table 21: Sample dataset of monthly workouts among 5 users

Let's first solve for S_D:

$$S_D = \sqrt{\frac{\Sigma x^2 - \frac{(\Sigma x)^2}{n}}{n-1}} = \sqrt{\frac{70 - \frac{(12)^2}{5}}{5-1}}$$

$S_D = \sqrt{(70 - 12^2 / 5) / (5 - 1)}$

$S_D = \sqrt{(70 - 144 / 5) / 4}$

$S_D = \sqrt{(70 - 28.8) / 4}$

$s_D = \sqrt{(41.2 / 4)}$

$s_D = \sqrt{10.3}$

$s_D = 3.209$

Let's now solve the full equation:

$$T = \frac{\bar{x}_D - \mu_0}{\frac{s_D}{\sqrt{n}}} \qquad T = \frac{2.4 - 0}{\frac{3.209}{\sqrt{5}}}$$

$\bar{x}_D = 2.4$ (mean of the difference: 12/5)

$\mu_0 = 0$ (the null hypothesis is no difference between μ_1 and μ_2)

$s_D = 3.209$ (standard deviation of the difference)

$n = 5$

$T = (2.4 - 0) / (3.209 / \sqrt{5})$

$T = 2.4 / (3.209 / 2.236)$

$T = 2.4 / 1.435$

T-Statistic = 1.672

Degrees of Freedom	Total area in one tail (a)					
	0.0005	0.001	0.005	0.010	0.025	0.050
1	636.6192	318.3088	63.65674	31.82052	12.70620	6.313752
2	31.59905	22.32712	9.924843	6.964557	4.302653	2.919986
3	12.92398	10.21453	5.840909	4.540703	3.182446	2.353363
4	8.610302	7.173182	4.604095	3.746947	2.776445	2.131847

Table 22: Excerpt of a T-distribution table

Based on 4 degrees of freedom with an alpha of 0.025 in each tail, the critical value is 2.776. Given our T-Statistic was 1.672, we can conclude the samples are not significantly different and therein accept the null hypothesis of $\mu_1 = \mu_2$.

3) One-sample T-Test

A one-sample T-Test is used for testing the sample mean of a single group against a known or hypothesized mean.

$$T = \frac{\bar{x} - \mu}{\dfrac{s}{\sqrt{n}}}$$

Where:

x = mean of the sample

μ = the known mean (null hypothesis)

s = standard deviation of the sample

n = sample size

We will run through an extended example of this T-Test method in the next chapter.

Summing Up

In conclusion, both the T-Test and Z-Test are used in hypothesis testing to decide whether you should accept or reject the null hypothesis based on a statistic. A T-Test is used in scenarios when you have a small sample size or you don't know the standard deviation of the population and you instead use T-distribution and the standard deviation of the sample. A Z-Test, meanwhile, is used for datasets with 30 or more observations (normal distribution) with a known standard deviation of the population and is calculated based on Z-distribution.

11

T-TEST CASE STUDY

Golf is played across flat terrain with nascent boundaries, but Andre Tolme pushed those boundaries when he completed an 18-hole course spanning the full length of a mid-sized country. Traversing 1,234 miles (or 1,985 kilometers) across the fairway and rough of the steppe, Tolme played his way from one side of Mongolia to the other over two summers. He completed the self-designed course on foot—without a golf buggy or even a local mare.

Tolme explains the guidelines of the bizarre challenge on his blog www.golfmongolia.com.

"I hit a golf ball as far as I can with a 3-iron, find it, then hit it again until I've covered more than 2,100 km (1,300 miles) from east to west across Mongolia...I have divided the country into 18 holes which each end in a town or city. Once a hole is finished, I pick up my ball and proceed to the far end of the urban area and tee it up again."[34]

For a totally fabricated experiment, let's suppose we challenge Andre's world record claim. Despite regular blog posts and pictures with an iron club in hand, it's not entirely unfeasible to suggest he faked it. In the book, *The Happiness of Pursuit*, author and world traveler Chris Guillebeau sheds light on the extreme measures a minority of people take in pursuing their

---

[34] Andre Tolme, "About Andre Tolme and Golf Mongolia," *Golf Mongolia*, accessed August 1, 2017, http://golfmongolia.com/About%20Golf%20Mongolia.html/.

bucket list of life goals. The author writes about a middle-aged dentist from Michigan who set out to run fifty marathons but was later discovered to have cheated along the way. This included a marathon held in West Wyoming (USA) that never happened but which even had its own website complete with the race times of fictitious runners!

Having heard the case of a dentist fabricating a city marathon, perhaps you now want to take a second look at Andre's scorecard.

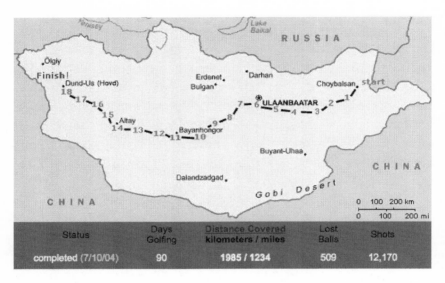

Status	Days Golfing	Distance Covered kilometers / miles	Lost Balls	Shots
completed (7/10/04)	90	1985 / 1234	509	12,170

Figure 18: Map of the 18-hole course, Source: Golf Mongolia, http://golfmongolia.com/

I do want to clarify that no one has publicly queried the validity of Andre Tolme's world record and which also helped to raise money for local charities. The content in this chapter is merely a hypothetical learning exercise and in no way undermining Andre's amazing achievement.

Let's begin by converting this problem into a hypothesis statement that incorporates a null hypothesis and an alternative hypothesis. Failure to include a null hypothesis in our research would be considered unprofessional and potentially invalid by the statistics and research community. Nullifying the null hypothesis is generally the most effective way to alter people's minds about a commonly accepted way of thinking. To validate an alternative hypothesis, it's therefore vital that we consider and include a null hypothesis as part of our experiment. Keep in mind, too, that failing to reject the null hypothesis doesn't validate the null hypothesis as true, only that the experiment did not find sufficient evidence to reject it.

The null hypothesis can be chosen from a variety of statements, but for now, let's contain it to the following:

"Andre set a world record by hitting golf balls across Mongolia with 12,170 shots."

In mathematical terms, this can be expressed as $H_0: \mu = 12{,}170$. The Greek character μ (mu) represents the population mean, which in this case is the result of the experiment. This means that our null hypothesis (H_0) predicts that our experiment's result will be equal to 12,170. Although we could add the feature of "90 days" to the hypothesis statement, this would make the experiment invalid, as this would be testing two variables ("shots" and "days").

There are also additional reasons why we shouldn't focus on the variable of time for this experiment. First, the timeframe could have been fabricated using Google Maps to produce a plausible number of days to cross Mongolia on foot and that would be difficult for the experiment to dispel. Second, Andre's expedition was not a time challenge; he did not set out to cross Mongolia in the shortest period of time possible or to break a previous time mark. Nor is golf a foot race, as any keen golfer will tell you.

So, while 90 days presents a tidy number for Andre to have completed the challenge, it's natural that he paced himself along the journey (the range of his recorded hits per day was 14 to 167). Also, by altering the amount of time spent looking for lost balls (of which he reported losing 509!) or spending less/more time on the fairway, he had considerable flexibility when it came to the completion date of his challenge.

His scorecard, though, is more likely to contain deception. The number of shots to complete the course is more difficult than *time* to predict and fabricate. As a person dedicated to the sport, it's also natural that Andre would want to record a low and competitive score to secure a Guinness World Record and uphold his reputation in the golf community. In summary, the number of times Andre hit the ball is the most likely variable to contain deception or foul play rather than any variable of time. This is therefore the variable we need to isolate and examine in our experiment.

Our Experiment

The fact that Andre completed the entire course with a 3-iron (a long club with a lofted angle face used for striking the ball 170 to 200 meters) offers useful guidance in designing our experiment. A full bag of clubs, each with unique advantages over terrain and distance, would make it complicated to replicate Andre's game over such a long distance. Using only a 3-iron, there's less variability regarding the ball's distance of movement from each hit, and we can make more reliable predictions over how many swings it might take to complete the course.

Finally, it's important to clarify that our experiment is limited to testing the validity of Andre's world record claim. We can't ascertain whether Andre rested for three days and allowed his caddie to swing for him, or whether he skipped swathes of the course on a four-by-four vehicle. Hence, our experiment is not a forensic or comprehensive assessment of whether Andre fabricated his scorecard but rather a simple study testing the plausibility of his score. Our study assumes that he did hit every ball with a 3-iron, that he did not skip parts of the course, and that he did not partake in any other form of behavior contrary to the spirit of the game.

Designing the Test

As we don't have the resources to attempt a 90-day course across Mongolia over a two-year period (Mongolia's extreme weather is not conducive for playing golf for most months of the year), we shall design our experiment around the first three holes of the course. There are three options for this experiment as outlined below.

1) An **independent samples T-Test** (two-sample test) where we ask Andre to replay the first three holes of the course and compare the mean against that of another sample of golfers (of similar playing ability). This would involve comparing the performance of two groups in the same experiment.

2) A **dependent sample T-Test** (paired test) where we ask Andre to replay the first three holes of the course and compare the mean with his previous results. This means measuring Andre's performance at different intervals.

3) A **one-sample T-Test** where we compare the known mean of Andre's previous results with that of a sample of other golfers (of similar playing ability).

Given that we already know the mean of Andre's previous scorecard, a one-sample T-Test is sufficient for this experiment. It would be problematic asking Andre to volunteer his time to participate in our experiment which rules out both an independent sample test and dependent sample test. Practical considerations are often important when deciding which type of test to use.

In terms of designing a one-sample T-Test, we will organize five subjects to play the first three holes of the course. To minimize performance discrepancies between our results and that of Andre's, we'll field five golfers of similar playing ability as Andre based on footage of his swing available on YouTube (https://www.youtube.com/watch?v=HdRBHWdKx-g).

Scorecard

(results in white)

Hole	1	2	3	4	5	6	7	8	9
yards	144,607	83,242	62,240	117,370	139,794	78,210	112,447	151,935	153,886
miles	82.2	47.3	35.4	66.7	79.4	44.4	63.9	86.3	87.3
km	132.2	76.1	56.9	107.3	127.8	71.5	102.8	138.9	140.5
Par	711	403	694	493	901	683	711	638	839
Shots	833	430	344	631	771	436	609	893	907
Lost Balls	40	23	18	34	43	28	35	72	59
Days	8	4	3	5.5	5.5	3	5	6	6

Hole	10	11	12	13	14	15	16	17	18	Totals
yards	147,122	72,631	140,669	196,783	97,899	132,574	113,213	136,075	91,008	2,171,505
miles	83.6	41.3	79.9	111.8	55.6	75.3	64.3	77.3	51.7	1,233.8
km	134.5	68.4	128.6	179.9	89.5	121.2	103.5	124.4	83.2	1985.2
Par	699	414	565	845	772	621	677	706	508	11,880
Shots	891	387	748	1096	527	778	627	756	506	12,170
Lost Balls	22	21	17	20	9	21	10	26	11	509
Days	7	3	5	7	4	5	4	5	4	90

Table 23: Andre's Scorecard, Source: Golf Mongolia, http://golfmongolia.com/Scorecard.html

Based on our experiment design of playing the first three holes, we also need to update our null hypothesis to:

"Andre completed the first three holes of the Mongolia course in 1607 shots." This is expressed as $H_0: \mu = 1607$.

One-tailed vs Two-tailed

As we know from the previous chapter, there are two ways to determine the validity of an experiment's results (one-tailed or two-tailed test). Currently, our null hypothesis is written as a two-tailed test because a score significantly lower or higher than $\mu = 1607$ is evidence to reject the null hypothesis. But for this experiment, we only need to examine one tail of the curve. The fact that the combined par (expected number of hits) for the first three holes was 1808 (711 + 403 + 694) is already evidence that a score higher than 1607 is plausible. Secondly, inflating his score would not be in Andre's personal interest. Hence, we only need to test the right-hand side of a one-tailed test.

If the experiment's results show that it's realistically possible for Andre to have completed the first three holes in 1607 or fewer shots, then the null hypothesis stands true. If the experiment delivers a result that is significantly higher than 1607, the alternative hypothesis is true, and hence, the rejection of the null hypothesis can only be accepted in one direction.

If our experiment finds that its feasible for Andre to have completed the hole in fewer than 1607 shots—say 1600 shots—then we can hardly dispel him of cheating unless he added shots to his scorecard to fool us. Regardless, we'll assume that Andre honestly recorded 1607 on his scorecard and we'll proceed with a right-hand side one-tail test to assess if such a score is possible.

Scorecard	
Golfer 1	1590
Golfer 2	1544
Golfer 3	1567
Golfer 4	1620
Golfer 5	1701
Mean	**1604.4**

Table 24: Experiment results

This table contains the results from our five golfers for the first three holes of the Mongolia course. Let's now compare the results with Andre's mean of 1607 using a one-sample T-Test.

$$T = \frac{\bar{x} - \mu}{\frac{s}{\sqrt{n}}}$$

x = 1604.4 (sample mean)

μ = 1607 (known mean)

s = 54.46 (standard deviation of sample)

n = 5 (size of sample)

T = (1604.4 − 1607) / (54.46 / $\sqrt{5}$)

T = -2.6 / (54.46 / 2.236)

T = -2.6 / 24.356

T-Statistic = -0.107

Using a T-distribution table, let's now look up the critical value based on an alpha of 0.05 and 4 degrees of freedom.

Degrees of Freedom	Total area in one tail (a)					
	0.001	0.005	0.010	0.025	0.050	0.100
1	318.3088	63.65674	31.82052	12.70620	6.313752	3.077684
2	22.32712	9.924843	6.964557	4.302653	2.919986	1.885618
3	10.21453	5.840909	4.540703	3.182446	2.353363	1.637744
4	7.173182	4.604095	3.746947	2.776445	2.131847	1.533206

Table 25: Excerpt of a T-distribution table

The intersection of df 4 and an alpha of 0.05 in one-tail is 2.131847, which means we can expect 95% of results to fall within 2.131847 standard deviations of the null hypothesis. As our sample's mean is -0.107, the results are not statistically significant and we do not have sufficient evidence to reject the null hypothesis.

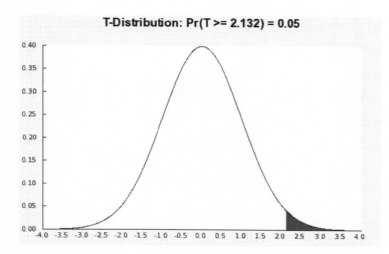

Figure 19: Right-hand T-Test with a critical value of 2.132

Based on this experiment, Andre's score for the first three holes of the Mongolia course seems humanly possible and we have no evidence to reject his world record. (Again, keep in mind this is merely an example of a hypothesis test experiment.)

12

COMPARING PROPORTIONS

In this chapter, we turn our attention to hypothesis testing for comparing two proportions from the same population expressed in percentage form, i.e. 40% of males vs 60% of females. This technique of comparing proportions is commonly used for analyzing experiments divided into an experimental group and a control group (placebo).

To compare two proportions, we need to conduct a two-proportion Z-Test using the following equation:

$$Z = \frac{(\hat{p}_1 - \hat{p}_2)}{\sqrt{\hat{p}(1 - \hat{p})\left(\frac{1}{n_1} + \frac{1}{n_2}\right)}}$$

Where:

p_1 = proportion 1

p_2 = proportion 2

p = proportion 1 and 2 combined

n_1 = sample size of proportion 1

n_2 = sample size of proportion 2

Exercise

In the following exercise, we consider a new energy drink formula that proposes to improve students' SAT scores.

The SAT *is a standardized test widely used for college admissions in the U.S. The maximum composite score for the test is currently 800 (reading/writing) + 800 (math) or 1600 in total. The average score is between 1050 and 1060 points.*

In our experiment, the benchmark for evaluating the students' results is based on whether they exceed 1,060 points on the test. The experiment considers a sample of 2,000 students split evenly into an experimental group and a control group. The experimental group is assigned the energy drink, and the control group receives the placebo which is a substance made to resemble the new energy drink but does not contain any active ingredients to enhance student performance.

From the two user groups, we receive the following results:

Experimental Group = 620/1000
Control Group = 500/1000

The number of students who surpassed 1,060 points on the test was 620 from the experimental group and 500 from the control group. Based on these results, it seems that the energy drink enabled the experimental group to surpass the average results of the control group.

Similar to how we used hypothesis testing to examine if a set of (golf) results were a likely or unlikely outcome, we are now going to compare two proportions to discern the likelihood that our new energy drink enhances students' SAT marks as the experiment's results suggest. This involves using hypothesis testing to compare the difference between the distributions to cancel out a random sample error manipulating the results and confirm that the experimental and control groups were evenly split. It's possible that the experimental group was assigned more above-average-students than the control group. If the groups were not split evenly, this might mean the experimental group was predetermined to surpass the control group irrespective of the effect of the new energy drink. To test the efficiency of our random sample, let's conduct a two-proportion Z-Test based on the following hypotheses.

$H_0: p_1 = p_2$
(The proportions are the same with the difference equal to 0)

$H_1: p_1 \neq p_2$
(The two proportions are not the same)

Note that we again anchored the null hypothesis with the statement that we wish to nullify. In this case, the null hypothesis is that the two proportions of results are identical and it just so happened that the results of the experimental group surpassed that of the control group due to a random sampling error. The null hypothesis therefore holds that the drug test did **not** have a substantial effect as the results could just as likely been reversed given a different random sample. However, based on the evidence so far, we actually want to attempt to nullify this hypothesis using a hypothesis test.

The next step is to define our confidence level, which again is 95%. This means that we'll reject the null hypothesis if there's a less than 5% chance of the alternative hypothesis occurring by chance. We can use a two-proportion Z-Test to perform the hypothesis test using the following equation to compare the two proportions.

$$\frac{(\hat{p}_1 - \hat{p}_2)}{\sqrt{\hat{p}(1 - \hat{p})\left(\frac{1}{n_1} + \frac{1}{n_2}\right)}}$$

Where:
$p_1 = 0.62$

$p_2 = 0.5$

p = 0.56 (both sample groups, 620+500/2000=0.56)

$n_1 = 1000$

$n_2 = 1000$

$$Z = \frac{(0.62 - 0.5)}{\sqrt{0.56(1-0.56)\left(\frac{1}{1000}+\frac{1}{1000}\right)}}$$

Z = (0.62 – 0.5) / √(0.56(1 – 0.56)(1/1000 + 1/1000))

Z = 0.12 / √(0.56(0.44)(2/1000))

Z = 0.12 / √(0.2464(0.002))

Z = 0.12 / √0.0004928

Z = 0.12 / 0.02219909908

Z-Statistic = 5.406

Let's review the normal distribution curve to visualize the results.

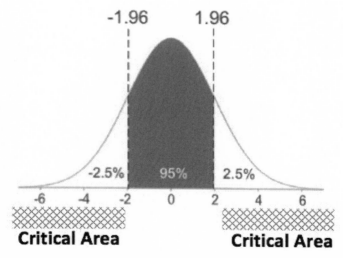

Figure 20: Normal distribution curve with marked critical areas

There are critical areas of 2.5% on each side of the two-tailed curve from a distance of 1.96 standard deviations. Thus, if the Z-Statistic falls within 1.96 standard deviations of the mean, we can conclude that the proportions of the experimental test and control test results were equal. Naturally, there will be some deviation from zero, but, of course, within 1.96 standard deviations using a 95% confidence level.

Given, though, that our actual Z-Statistic is 5.4 standard deviations from the mean, and higher than 1.96, we must therefore reject the null hypothesis. This means that it's doubtful that the results of our experiment occurred by random chance and there was a variable at play stronger than a poor sampling split, which in this case, is our new and commercially promising energy drink.

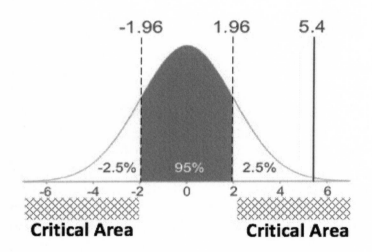

Figure 21: Results of the hypothesis test added to the curve

13

REGRESSION & CORRELATION

A useful and widely practiced technique in inferential statistics is regression analysis, which is used to test how well a variable predicts another variable.

The term "regression" is derived from Latin, meaning "going back," and was popularized by Francis Galton in the late 19th Century. In his paper *Regression towards Mediocrity in Hereditary Stature*, Galton observed that extreme characteristics such as the height of your parents isn't always a determinate of height. Galton conducted multiple experiments and in one study he measured the heights of parents and adjusted for the height of females by multiplying their height by a weight of 1.08. The average height of the two parents was calculated to produce a "midheight" that was compared with the height of the parents' adult child.

Galton's study found that the parents' midheight had a mean of 68.2 inches (173cm) and while some parents in the study recorded heights below and above the mean score, their child's height tended to regress (be closer) to the average height of all people in the study rather than that of the parents. This helps to explain why extraordinarily tall or short people don't necessarily share the same height as their children. Galton named this phenomenon "regression towards mediocrity." Today, it is expressed more affirmatively as "regression towards the mean."

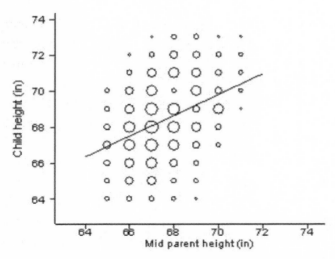

Figure 22: Plotted results of Galton's experiment, Source: Martin Bland, University of York

This two-dimensional scatterplot is based on Galton's original data and shows parents' midheight recorded in inches on the horizontal axis and their child's height on the vertical axis. The size of the circular data points on the scatterplot captures the number of reoccurring instances, with larger circles representing common data points.

Data points are created by inputting their respected values on the x and y-axes. Dependent variables are plotted on the y-axis and independent variables on the x-axis. Dependent variables (y) denote what you're attempting to predict, which in this example is the child's height. The independent variable (X) is the variable that supposedly impacts the dependent variable (y), which is the parents' midheight.

The objective of regression analysis is to find a line that best fits the data points on the scatterplot to make predictions. In the case of linear regression, the line is straight and cannot curve or pivot. Nonlinear regression, meanwhile, grants the line to curve and bend to fit the data.

As a straight line cannot possibly intercept all data points, linear regression can be thought of as a trendline (which is how it's often referred to in software programs like Google Sheets) visualizing the underlying trend of the dataset. This means that if you were to draw a perpendicular line (at an angle of 90 degrees) from the regression line (known as the hyperplane) to each data point on the scatterplot, the aggregate distance of each point would equate to the smallest possible distance to the hyperplane.

Another important feature of regression is the *slope,* also often known as the *coefficient* in statistics. The term "coefficient" is generally used over "slope" in cases where there are multiple variables in the equation (multiple linear regression) and the line's slope is not explained by any single variable.

The slope can be found by referencing the hyperplane; as one variable increases, the other variable increases by the average value denoted by the hyperplane. The slope is therefore useful in forming predictions. We can use the slope, for example, to find the intercept between parents' midheight (X) of 72 inches and their adult child's expected height (y). In this case, the y value is approximately 71 inches.

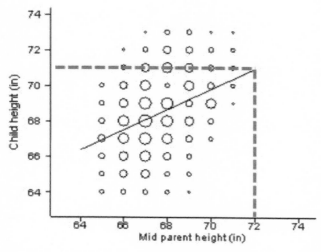

Figure 23: Predicted height of a child whose parents' midheight is 72 inches

Regression towards the mean or regression analysis is thus a useful method for estimating relationships among variables and testing if they're somehow related.

While linear regression is not a fail-proof method of making predictions, the trendline does offer a primary reference point to make estimates about the future. Nonlinear regression can also be used to optimize the model by modifying the slope to better fit the data.

Learning the Formula

The mathematical formula for linear regression used in statistics is $y = bx + a$. (Although the linear equation is written differently in other disciplines, such as $y = ax + b$, this is the preferred style used in statistics.) The "y" represents the dependent variable, and "x" represents the independent variable. The variable "a" represents the point where the hyperplane crosses the y-axis, known as the *y-intercept*, and the "b" represents the hyperplane's slope and dictates its steepness. The variable "a" or y-intercept is the value of y when $x = 0$, and "b" or slope explains the relationship between x and y (what change in y is predicted for 1 unit change in x?). If "b" is positive, the regression line (in linear cases) rises from left to right. If "b" is negative, the regression line falls from left to right. If "b" is zero, the line is flat or horizontal.

We'll now run through an example using the following dataset, which is free to download from Kaggle.com.[35] The dataset contains a list of video games with sales higher than 100,000 copies. The following table is a small excerpt of the full dataset combined with my own calculations shown in columns 4 and 5.

---

[35] Kendall Gillies, "Video Game Sales and Ratings," *Kaggle*, accessed August 25, 2017, https://www.kaggle.com/kendallgillies/video-game-sales-and-ratings/data/.

Video Game Name	Japan Sales (x)	Global Sales (y)	xy	x^2
Wii Sports	3.77	82.74	311.9298	14.2129
Super Mario Bros	6.81	40.24	274.0344	46.3761
Mario Kart	3.79	35.82	135.7578	14.3641
Wii Sports Resort	3.28	33	108.24	10.7584
Pokemon Red/Pokemon Blue	10.22	31.37	320.6014	104.4484
Tetris	4.22	30.26	127.6972	17.8084
New Super Mario Bros (2006)	6.5	30.01	195.065	42.25
Wii Play	2.93	29.02	85.0286	8.5849
New Super Mario Bros (2009)	4.7	28.62	134.514	22.09
Duck Hunt	0.28	28.31	7.9268	0.0784
Σ (Total Sum)	46.5	369.39	1700.795	280.9716

Table 26: Video game sales for Japan and Global markets

Formula for Linear Regression

$$a = \frac{(\Sigma y)(\Sigma x^2) - (\Sigma x)(\Sigma xy)}{n(\Sigma x^2) - (\Sigma x)^2}$$

$$b = \frac{n(\Sigma xy) - (\Sigma x)(\Sigma y)}{n(\Sigma x^2) - (\Sigma x)^2}$$

With our dataset and formula available in front of us, we can complete the equation by plugging in the values from the data.

Remember that "**Σ**" equals the total sum. So Σxy is the total sum of x multiplied by y. Also, "**n**" equals the total number of items, which in this case is 10 (as there are 10 Nintendo Games in the dataset).

Σy = 369.39

Σx = 46.5

Σxy = 1700.795

Σx² = 280.9716

n = 10

STEP 1

Find the value of "a" (the y-intercept):

((369.39 × 280.9716) – (46.5 × 1700.795)) / (10(280.9716) – 46.5^2)

(103788.09932 – 79086.9675) / (2809.716 – 2162.25)

24701.13182 / 647.466

= **38.15**

STEP 2

Find the value of "b" (the slope):

(10(1700.795) – (46.5 × 369.39)) / (10(280.9716) – 46.5^2)

(17007.95 – 17176.635) / (2809.716 – 2162.25)

-168.685 / 647.466

= **-0.2605**

STEP 3

Insert the "a" and "b" values into a linear equation.

y = bx + a
y = -0.2605x + 38.15

The linear equation y = -0.2605x + 38.15 sets the location of the regression hyperplane, with 38.15 as the y-intercept and a negative slope of -0.2605.[36]

---

[36] This equation could also be expressed using the notation of $y = \beta_0 + \beta_1 x_1 + e$, where β_0 is the intercept, β_1 is the slope, and e is the residual or error (explained later in this chapter).

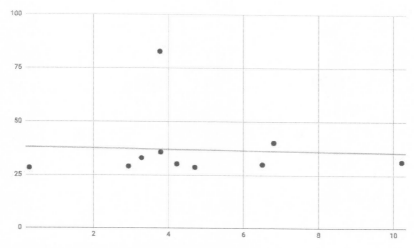

Figure 24: Nintendo sales plotted on a scatterplot

For both examples used in this chapter (height and Nintendo games), the trendline visibly approximates the underlying pattern in the data. This is because a small data sample enables us to crosscheck for accuracy using our visual senses. However, for large datasets with a higher degree of scattering or three-dimensional and four-dimensional data, it may not be as easy to validate the trendline as an optimal fit by looking at the trendline.

A mathematical solution to this problem is to apply *R-squared*, also known as the *coefficient of determination*, which defines the percentage of variance in the linear model in relation to the independent variable. In other words, R-squared is a test to see what level of impact the independent variable has on data variance.

Specifically, R-squared constitutes a number between "0" and "1," which produces a percentage value between 0% and 100%.

- 0% holds that the linear regression model accounts for none of the data variability in relation to the mean of the dataset. This indicates that the regression line is a poor fit for the given dataset.

- 100% holds that the linear regression model expresses all the data variability in relation to the mean of the dataset. This indicates that the regression line is a perfect fit for the dataset.

R-squared is calculated as the sum of square regression (SSR) divided by the sum of squares total (SST).

We'll now revisit the dataset for Nintendo games (Table 24) to learn how to calculate SSR and SST.

Video Game	x	y	y' (-0.2605x + 38.15)	SSR	SST
Wii Sports	3.77	82.74	37.167915	0.05240208	2097.7316
Super Mario Bros	6.81	40.24	36.375995	0.31697463	10.896601
Mario Kart	3.79	35.82	37.162705	0.05004393	1.252161
Wii Sports Resort	3.28	33	37.29556	0.12713503	15.515721
Pokemon Red/Blue	10.22	31.37	35.48769	2.10630072	31.013761
Tetris	4.22	30.26	37.05069	0.01247466	44.609041
New Super Mario Bros (2006)	6.5	30.01	36.45675	0.23256506	48.011041
Wii Play 2006 Nintendo	2.93	29.02	37.386735	0.20046663	62.710561
New Super Mario Bros (2009)	4.7	28.62	36.92565	0.00017822	69.205761
Duck Hunt	0.28	28.31	38.07706	1.29518056	74.459641
Σ Total Sum	46.5	369.39	N/A	4.39372152	2455.40589
Mean	4.65	36.939	N/A	N/A	N/A

Table 27: Video game sales for Japan (x) and Global (y) markets

As you may have noticed, three new columns have been added to the table, which I'll explain before we proceed to the final calculations.

Column 4 contains row-by-row results from the linear equation **y = -0.2605x + 38.15** that we generated in the previous exercise. This means that for each row, the x value shown in Column 2 is inputted into the linear equation to make an expected prediction (y').

Thus, for Wii Sports (Row 1) the outcome of Column 4 (y' = -0.2605x + 38.15) is calculated as:

y' = -0.2605(**3.77**) + 38.15

y' = -0.982085 + 38.15

y' = 37.167915

This process is repeated for each game, and in each case, the "x" value in the formula is determined by the "x" value contained in Column 2 from that same row.

Column 5 contains the SSR value for each video game. To calculate SSR, the predicted y value generated by the linear equation in Column 4 is added to the following equation as y' (predicted y value).

$$SSR = (y' - \bar{y})^2$$

Thus for Wii Sports, SSR is calculated as:

(predicted y value – mean y value)^2

$(37.167915 - 36.939)^2$

$(0.228915)^2 = 0.05240208$

This process is repeated for each Nintendo game listed in the table. For each row, the y' value is determined by the predicted y value contained in Column 4 in that row. The value of all ten rows is then added to produce a sum of square regression (SSR), which returns the value of 4.39372152.

Column 6 comprises the SST value of each row. To calculate the SST value for each video game, the mean y value is subtracted from the actual y value (Column 3) for the game and squared.

$$SST = (y - \bar{y})^2$$

Thus for Wii Sports, SST is calculated as:

(actual y value – mean y value)^2

$(82.74 - 36.939)^2$

$(45.801)^2 = \textbf{2097.7316}$

SST for Wii Sports produces a result of 2097.7316.

Again, this process is repeated for each game. The value of all ten rows is then added to produce a sum of squares total (SST), which is 2455.40589.

Lastly, we divide the sum of square regression (SSR) by the sum of squares total (SST) to generate R-squared.

R-squared = SSR / SST

R-squared = 4.39372152 / 2455.40589

R-squared = 0.0018

An R-squared statistic close to 1 signifies that a large proportion of the variability is captured by the regression model, whereas an R-squared statistic close to 0 indicates an error or a poor fit. Given our R-squared statistic of 0.0018 is close to 0 (bad fit), we can confirm that our regression line is not a close fit for our dataset, which is caused by the anomaly Wii Sports and its high y value (82.72).

Consideration of domain use can also play a role in interpreting the R-squared statistic. In *An Introduction to Statistical Learning with Applications in R*, the authors contrast a physics problem with a relatively small drop-off from 1 and a marketing experiment with a considerably lower R-squared value.[37] A modest drop-off in the field of physics might indicate a problem with the experiment, whereas a lower R-squared value might be tolerated in marketing experiments given the influence of unmeasured variables and other random phenomena that produce a rough approximation.

Pearson Correlation

Using linear regression, we found the equation ($y = bx + a$) for plotting the linear relationship between two variables and

---

[37] Gareth James, Daniela Witten & Trevor Hastie Robert Tibshirani, "An Introduction to Statistical Learning with Applications in R," *Springer*, 2017.

predicting the unit of change of the dependent variables based on the known value of the independent variable. Another common measure of association between two variables is the *Pearson correlation coefficient*, which describes the strength or absence of a relationship between two variables. This is slightly different from linear regression analysis, which expresses the average mathematical relationship between two or more variables with the intention of visually plotting the relationship on a scatterplot. Pearson correlation, meanwhile, is a statistical measure of the co-relationship between two variables without any designation to independent and dependent qualities.

Pearson correlation (r) is expressed as a number (coefficient) between -1 and 1. A coefficient of -1 denotes the existence of a strong negative correlation, 0 equates to no correlation, and 1 for a strong positive correlation. Thus, a correlation coefficient of -1 means that for every positive increase in one variable, there is a negative decrease of a fixed proportion in the variable, such as airplane fuel which decreases in line with distance flown. Conversely, a correlation coefficient of 1 signifies an equivalent positive increase in one variable based on a positive increase in another variable, such as food calories of a particular food that goes up with its serving size. Lastly, a correlation coefficient of zero notes that for every increase in one variable, there is neither a positive or negative change, which means the two variables aren't related.

According to the following entities, Pearson correlation coefficients can be roughly interpreted as follows.

Correlation Coefficient		Dancey & Reidy (Psychology)	Quinnipiac University (Politics)	Chan YH (Medicine)
+1	−1	Perfect	Perfect	Perfect
+0.9	−0.9	Strong	Very Strong	Very Strong
+0.8	−0.8	Strong	Very Strong	Very Strong
+0.7	−0.7	Strong	Very Strong	Moderate
+0.6	−0.6	Moderate	Strong	Moderate
+0.5	−0.5	Moderate	Strong	Fair
+0.4	−0.4	Moderate	Strong	Fair
+0.3	−0.3	Weak	Moderate	Fair
+0.2	−0.2	Weak	Weak	Poor
+0.1	−0.1	Weak	Negligible	Poor
0	0	Zero	None	None

Table 28: Interpretations of Pearson correlation coefficients. Source: Science Direct

The Pearson's correlation coefficient can be calculated using the following equation, which partially resembles the equation for b (slope) within linear regression (except with a different denominator).

$$r = \frac{n(\sum xy) - (\sum x)(\sum y)}{\sqrt{[n(\sum x^2) - (\sum x)^2][n(\sum y^2) - (\sum y)^2]}}$$

Video Game	Japan Sales (x)	Global Sales (y)	xy	x^2	y^2
Wii Sports	3.77	82.74	311.9298	14.2129	6845.9076
Super Mario Bros	6.81	40.24	274.0344	46.3761	1619.2576
Mario Kart	3.79	35.82	135.7578	14.3641	1283.0724
Wii Sports Resort	3.28	33	108.24	10.7584	1089
Pokemon Red/Pokemon Blue	10.22	31.37	320.6014	104.4484	984.0769
Tetris	4.22	30.26	127.6972	17.8084	915.6676
Sum (Σ)	32.09	253.43	1278.2606	207.9683	12736.9821

Table 29: Sample data

Calculations

Using the sample data (n=6) above, we can plug in each of the values into the Pearson's correlation coefficient equation shown below.

$r = (n(\Sigma xy) - (\Sigma x)(\Sigma y)) / \sqrt{(n(\Sigma x^2) - (\Sigma x)^2)(n(\Sigma y^2) - (\Sigma y)^2)}$

$r = (6(1278.2606) - (32.09)(253.43)) / \sqrt{(6(207.9683) - (32.09)^2)(6(12736.9821) - (253.43)^2)}$

$r = (7669.5636 - 8132.5687) / \sqrt{((1247.8098 - 1029.7681)(76421.8926 - 64226.7649))}$

$r = -463.0051 / \sqrt{((218.0417)(12195.1277))}$

$r = -463.0051 / \sqrt{2659046.37542509}$

$r = -463.0051 / 1630.65826445$

$r = -0.284$

Japan Sales for our sample data therefore has a weak negative correlation score of -0.284 with Global Sales, which means that for every positive increase in Japan Sales (x), there is a weak negative decrease of a fixed proportion in Global Sales (y). This negative Pearson correlation coefficient, though, is largely exaggerated by the outlier score from the first data point (Wii Sports) which has an abnormally high y value (82.74).

Testing Statistical Significance of Pearson Correlation

The correlation coefficient can also be used for hypothesis testing to dispel the null hypothesis. The null hypothesis can be set to r = 0 and the alternative hypothesis to r ≠ 0.

In the following example we'll test if the sample outcome is significantly different from 0 using a significance test with an alpha of 0.05 and Pearson correlation of 0.81. The formula for the significant test for the Pearson correlation coefficient with a t-distribution is calculated with n−2 degrees of freedom (n is the sample size), as shown here.

$$T = \frac{r\sqrt{n-2}}{\sqrt{1-r^2}}$$

Sample Data

n = 18 (sample size)

r = 0.81 (Pearson correlation between two given variables)

t = (0.81√(18 − 2)) / √(1 − 0.81²)

t = 0.81(√16) / √(1 − 0.6561)

t = 0.81(4) / √0.3439

t = 3.24 / 0.586

t = 5.529

After calculating the T-Score for the correlation between our two sample variables, we can compare this result with the critical area value for our experiment. Using a T-distribution table, based on a two-tailed T-Test with 16 degrees of freedom and an alpha of 0.05 (0.025 in each tail), the critical area value is equal to 2.119905.

Degrees of Freedom	Total area in one tail (a)					
	0.0005	0.001	0.005	0.010	0.025	0.050
15	4.072765	3.732834	2.946713	2.602480	2.131450	1.753050
16	4.014996	3.686155	2.920782	2.583487	2.119905	1.745884

Table 30: Excerpt of a T-distribution table

Given that our computed T-Score of 5.529 is greater than the critical area value of 2.119905, we can thereby reject the null hypothesis that these two sample variables are unrelated.

14

CLUSTERING ANALYSIS

Having analyzed relationships between variables as data points plotted on a scatterplot in the previous chapter, let's look at a technique to group these data points called clustering analysis.

As a statistical-based method of data segmentation, clustering analysis aims to group similar objects (data points) into clusters based on the chosen variables. This method partitions data into assigned segments or subsets where objects in one cluster resemble one another and are dissimilar to objects contained in the other cluster(s). Objects can be interval, ordinal, continuous or categorical variables. However, a mixture of different variable types can lead to complications with your analysis because the measures of distance between objects can vary depending on the variable types contained in the data.

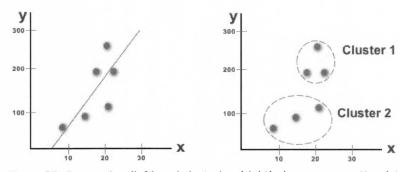

Figure 25: Regression (left) and clustering (right) shown on a scatterplot

Clustering analysis developed originally from anthropology in 1932, before it was introduced to psychology in 1938 and was later adopted by personality psychology in 1943 for trait theory classification. Today, clustering analysis is used in data mining, information retrieval, machine learning, text mining, web analysis, marketing, medical diagnosis, and numerous other fields. Specific use cases include analyzing symptoms, identifying clusters of similar genes, segmenting communities in ecology, and identifying objects in images.

Clustering analysis, however, is not one fixed technique but rather a family of methods, which includes hierarchical clustering analysis and non-hierarchical clustering as we will explore next.

Hierarchical Clustering Analysis

Hierarchical clustering analysis (HCA) is a technique to build a hierarchy of clusters. An example of a HCA approach is divisive hierarchical clustering, which is a top-down method where all objects start as a single cluster and are split into pairs of clusters until each object represents an individual cluster.

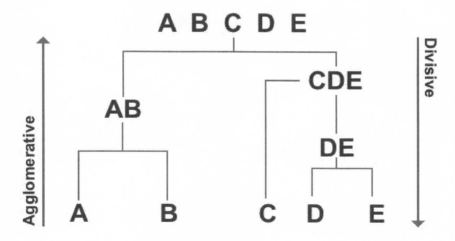

Figure 26: Hierarchical Cluster Analysis

The alternative and more popular approach is agglomerative hierarchical clustering, which is a bottom-up method of classification. Opposite to the divisive method, agglomerative clustering is carried out in reverse, where each object starts as a standalone cluster, and a hierarchy is created by merging pairs of clusters to form progressively larger clusters. This technique can be broken down into the following three steps:

1) Objects start as their own separate cluster, which results in a maximum number of clusters.

2) The number of clusters is reduced by combining the two nearest (most similar) clusters. Note that agglomerative clustering methods differentiate by the interpretation of the "shortest distance," which we'll examine shortly.

3) This process is repeated until all objects are grouped inside one single cluster.

As a result, hierarchical clusters resemble a series of nested clusters organized within a hierarchical tree. The agglomerate cluster starts with a broad base and a maximum number of clusters. The number of clusters falls at subsequent rounds until there's one single cluster at the top of the tree.

In the case of divisive clustering, the tree is upside down. At the bottom of the tree is one single cluster that contains multiple loosely related clusters. These clusters are sequentially split into smaller clusters until the maximum number of clusters is reached.

Hierarchical clustering is commonly used in conjunction with a dendrogram chart to visualize the arrangement of clusters. Dendrogram charts are used to demonstrate taxonomic relationships and are commonly used in biology to map clusters of genes or other samples. The word itself is derived from the Greek word dendron, meaning "tree."

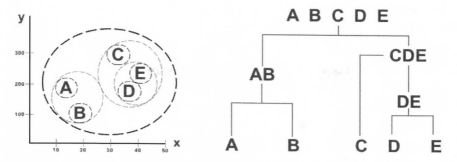

Figure 27: Nearest neighbor (left) and a hierarchical dendrogram (right)

Agglomerative Clustering Techniques

There are various methods to perform agglomerative hierarchical clustering analysis, and they differ in both the technique they use to find the "shortest distance" between clusters and in the shape of the clusters they produce.

Nearest Neighbor

Nearest neighbor is a well-known and straightforward method that creates clusters based on the distance between the two closest neighbors. Using this technique, you find the shortest distance between two objects and combine them into one cluster. This step is repeated so that the next shortest distance between two objects is found, which either expands the size of the first cluster or forms a new cluster between two objects.

Furthest Neighbor Method

The furthest neighbor method produces clusters by measuring the distance between the most distant pair of objects. Under this method, the distance between each possible object pair is computed, and the object pairs located furthest apart are unable to be linked. Meanwhile, at each stage of hierarchical clustering, the two closest objects are merged into a single cluster. Note that this method is relatively sensitive to outliers.

Average

Known also as UPGMA (Unweighted Pair Group Method with Arithmetic Mean), this clustering method merges objects by calculating the distance between two clusters and measuring the average distance between all objects in each cluster and joining the closest cluster pair. Initially, this technique is no different to nearest neighbors because the first cluster to be linked contains only one object. However, once a cluster includes two or more objects, the average distance between objects within the cluster can be measured which has an impact on classification.

Centroid Method

The centroid method utilizes the object in the center of each cluster, called the centroid, to determine the distance between two clusters. At each step, the two clusters whose centroids are measured to be closest together are merged.

Ward's Method

This method draws on the sum of squares error (SSE) between two clusters over all variables to determine the distance between clusters. Using Ward's method, all possible cluster pairs are combined, and the sum of the squared distance across all clusters is calculated. At each round, Ward's method attempts to merge two separate clusters by combining the two clusters that best minimize SSE. In other words, the pair of clusters that return the highest sum of squares is selected and conjoined.

This method produces clusters relatively equal in size, which may not always be effective. Another weakness of this method is that it can be sensitive to outliers. Nonetheless, it remains one of the most popular agglomerative clustering methods in use today.

Measures of Distance

In addition to the technique selected to assign clusters, your choice of measurement also has an impact on cluster composition.

This can lead to different classification results, as two objects may be positioned nearer than another pair of objects by one measurement but farther away according to another.

	Maximum	Euclidean	Manhattan
Distance	2	$\sqrt{1}$	1

Table 31: Distance between 1,0 and the origin of 0,0

Distance measurements vary according to their use application, but Euclidean distance tends to be the standard across most industries, including machine learning and psychology. Other popular variants include squared Euclidean distance, Manhattan distance (reduces the influence of outliers and resembles walking a city block), maximum distance, and Mahalanobis distance (internal cluster distances tend to be emphasized, whereas distances between clusters are less significant).

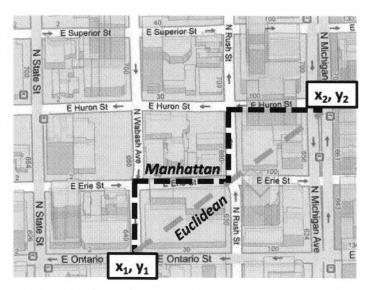

Figure 28: Manhattan distance versus Euclidean distance

Euclidean distance is found as follows:

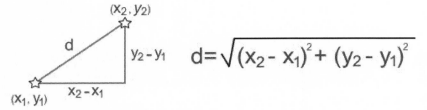

Figure 29: Euclidean distance formula

Nearest Neighbor Exercise

In the following practice exercise, we'll use nearest neighbor clustering to find two sections of the Mongolia course (discussed in Chapter 11) that are most similar based on two variables: par score and Andre's score. To keep this exercise simple, we'll examine only the first seven holes of the course and not the full 18 holes.

	Hole 1	Hole 2	Hole 3	Hole 4	Hole 5	Hole 6	Hole 7
Par (x)	711	403	694	493	901	683	711
Andre's Score (y)	891	430	344	631	771	436	609

Table 32: The anticipated par score (x) and Andre Tolme's actual score (y)

To find the two nearest neighbors based on our chosen variables, we'll calculate the Euclidean distance between each two-hole combination. To find the Euclidean distance, we'll need to use the following formula: $\sqrt{[(x_2 - x_1)^2 + (y_2 - y_1)^2]}$

Example: Hole 1 (H1) and Hole 2 (H2)

$\sqrt{[(\text{H1 Par} - \text{H2 Par})^2 + (\text{H1 Andre's Score} - \text{H2 Andre's Score})^2]}$

$\sqrt{[(711 - 403)^2 + (891 - 430)^2]}$

$\sqrt{[(308)^2 + (461)^2]}$

$\sqrt{[94,864 + 212,521]}$

$\sqrt{307{,}385}$

Euclidean distance = 554.42312

Example: Hole 1 (H1) and Hole 3 (H3)

$\sqrt{[(\text{H1 Par} - \text{H3 Par})^2 + (\text{H1 Andre's Score} - \text{H3 Andre's Score})^2]}$

$\sqrt{[(711 - 694)^2 + (891 - 344)^2]}$

$\sqrt{[(17)^2 + (547)^2]}$

$\sqrt{[289 + 299{,}209]}$

$\sqrt{299{,}498}$

Euclidean distance = 547.264

To calculate the Euclidean distance in Microsoft Excel, you can use the following formula (fx): =SQRT((B2-C2)^2 + (B3-C3)^2) as shown in the snippet below and by inputting the appropriate cell coordinates.

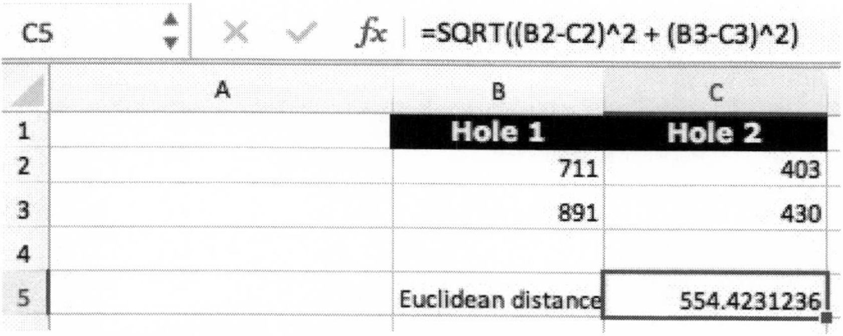

CS		fx =SQRT((B2-C2)^2 + (B3-C3)^2)	
	A	B	C
1		**Hole 1**	**Hole 2**
2		711	403
3		891	430
4			
5		Euclidean distance	554.4231236

Figure 30: Microsoft Excel function for calculating Euclidean distance

Let's now add the other calculations to the following utility matrix.

	Hole 1	Hole 2	Hole 3	Hole 4	Hole 5	Hole 6	Hole 7
Hole 1	NA	554.423	547.264	339.299	224.722	455.861	282
Hole 2		NA	303.442	220.230	603.560	280.064	356.237
Hole 3			NA	350.385	474.529	92.655	265.545
Hole 4				NA	431.351	272.259	219.107
Hole 5					NA	399.686	249.688
Hole 6						NA	175.251
Hole 7							NA

Table 33: Utility matrix of Euclidean distance between holes based on the two measured variables of x and y

First, you'll notice half the matrix is left incomplete, and this is deliberate because the matrix is symmetric. This means that each of the missing values can be found in the opposite corresponding cell inside the utility matrix and there's no need to complete the full matrix with every calculation.

From these calculations, we can see that Hole 3 and Hole 6 are the two nearest neighbors, and therefore the most similar, based on their Euclidean distance of 92.655. Conversely, the most dissimilar data points are Hole 2 and Hole 5 where the Euclidean distance is 603.560.

Finally, it's important to note that this method is computationally expensive as each combination of data points must be calculated before we can be certain that we've found the two closest neighbors.

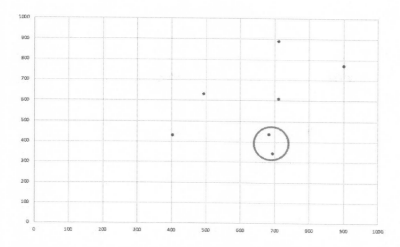

Figure 31: Data points plotted on a scatterplot & two nearest neighbors highlighted

Non-Hierarchical Clustering

Non-hierarchical or partitional clustering is different from hierarchical clustering and is commonly used in business analytics. Rather than nesting clusters inside large clusters, non-hierarchical methods divide *n* number of objects into *m* number of clusters. Also, unlike hierarchical clustering, each object can only be assigned to one cluster and each cluster is discrete. This means that there's no overlap between clusters and no case of nesting a cluster inside another.

For this reason, non-hierarchical methods are usually faster and require less storage space than hierarchical methods, which explains why they're typically used in business scenarios. Secondly, rather than mapping the hierarchy of relationships within a dataset using a dendrogram chart, non-hierarchical clustering helps to select the optimal number of clusters to perform classification.

HIERARCHICAL   NON-HIERARCHICAL

 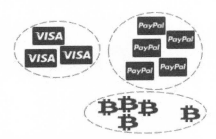

Figure 32: Hierarchical and non-hierarchical clustering for analysis of payments made on an online platform

Synonymous with non-hierarchical clustering is the *k*-means clustering technique, which is used prominently in machine learning and data mining as a classifier to predict unknown data points. The *k*-means clustering technique attempts to split data into *k* number of clusters, where *k* represents the number of clusters you wish to define. Setting *k* to "3," for example, splits the data into three clusters.

To split data into clusters, each cluster is assigned a centroid, which is a data point that forms the epicenter of an individual cluster. Centroids can be chosen at random, which means you can nominate any data point on the scatterplot to assume the centroid role. The remaining data points on the scatterplot are then assigned to the closest centroid by measuring the Euclidean distance.

After all data points have been allocated to a centroid, the next step is to aggregate the mean value of all data points for each cluster, which can be found by calculating the average x and y values of all data points in that cluster.

Next, take the mean value of the data points in each cluster and plug in those x and y values to update your centroid coordinates. This will most likely result in a change to your centroids' location. Your total number of clusters, however, will remain the same. You are not creating new clusters, but instead updating their position on the scatterplot. Like musical chairs, the remaining data points rush to the closest centroid to form *k* number of clusters. Should any data point on the scatterplot switch clusters with the changing of centroids, the previous step is repeated. This means, again, calculating the average mean value of the cluster and updating the x and y values of each centroid to reflect the average coordinates of the data points in that cluster.

Once you reach a stage where the data points no longer switch clusters after an update in centroid coordinates, the algorithm is complete, and you have your final set of clusters.

Below is a practical example of *k*-means clustering. The dataset for this example is comprised of seven bottles of beer (each a different brand) along with two variables: wholesale (cost) price and retail price. In this example, *k* is set to 2 to split the dataset into two clusters.

Bottle	Cost Price ($)	Retail Price ($)
A	1	2
B	3	5
C	5	6
D	5	7
E	2.5	3.5
F	5	8
G	3	4

Table 34: Cost and retail price of each bottle

Step 1: Let's first visualize this dataset on a scatterplot.

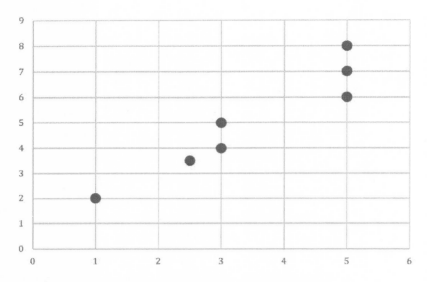

Figure 33: Sample data points are plotted on a scatterplot

Each data point on the scatterplot represents one beer bottle, with the horizontal x-axis representing wholesale price and the vertical y-axis representing retail price.

Step 2: With *k* set to 2, the next step is to split the data into two clusters by first nominating two data points to act as centroids. You can think of a centroid as a team leader. Other data points then report to the closest centroid according to their location on the scatterplot. Centroids can be chosen at random, and in this example, we have nominated data points A (1, 2) and D (5, 7) to act as our two initial centroids. The two centroids are now represented as larger circles on the scatterplot.

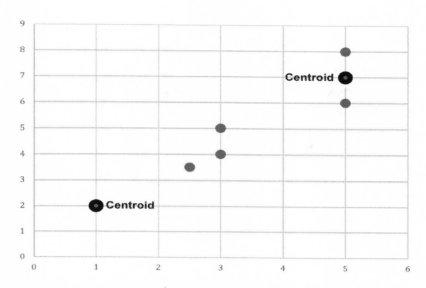

Figure 34: Two existing data points are nominated as the centroids

Step 3: The remaining data points are assigned to the closest centroid, as summarized in the following table.

Cluster 1		Cluster 2	
Bottle	**Mean Value**	**Bottle**	**Mean Value**
A* (1.0, 2.0)	(1.0, 2.0)	D* (5.0, 7.0)	(5.0, 7.0)
		B (3.0, 5.0)	(4.0, 6.0)
		C (5.0, 6.0)	(4.33, 6.0)
E (2.5, 3.5)	(1.75, 2.75)		
		F (5.0, 8.0)	(4.5, 6.5)
G (3.0, 4.0)	(2.16, 3.2)		
A, E, G	**(2.16, 3.2)**	**D, B, C, F**	**(4.5, 6.5)**

Table 35: Cluster 1 and 2 coordinates
* Centroid

Cluster 1 comprises data points A, E, G, and together their mean coordinates are **2.16**, **3.2**. Cluster 2 comprises data points B, C, D, F, and together their mean coordinates are **4.5**, **6.5**. The two clusters and their respective centroids (A and D) are visualized in Figure 35.

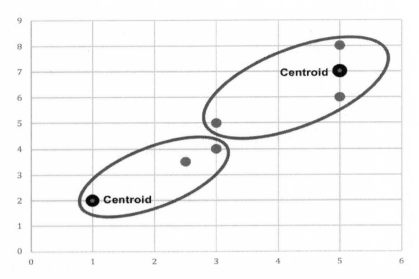

Figure 35: Two clusters are visualized after calculating the Euclidean distance of the remaining data points to the centroids

Step 4: We now use the mean coordinates, calculated in Table 32, to update our two centroids' location. The two previous centroids stay in their original position and two new centroids are added to the scatterplot. The new centroid location for Cluster 1 is **2.16, 3.2** and the new centroid location for Cluster 2 is **4.5, 6.5**.

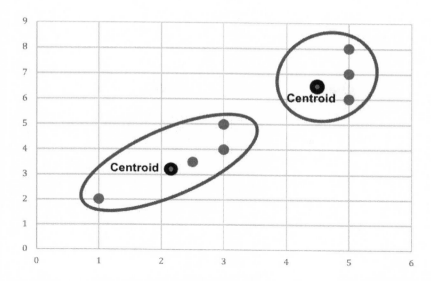

Figure 36: The centroid coordinates for each cluster are updated to reflect the cluster's mean coordinates. As one data point has switched from the right cluster to the left cluster, the centroids of both clusters need to be updated one last time.

Step 5: Next, we need to check that each data point remains aligned with its updated centroid. Immediately we see that one data point has switched sides and joined the opposite cluster! That data point is B (3, 5). We thus need to go back and update the mean value of each cluster, with data point B now assigned as a group member of Cluster 1, rather than Cluster 2.

Cluster 1			Cluster 2		
	X Value	**Y Value**		**X Value**	**Y Value**
A	1	2	C	5	6
B	3	5	D	5	7
E	2.5	3.5	F	5	8
G	3	4			
Mean	**2.4**	**3.5**	**Mean**	5	7

Table 36: Cluster 1 and 2 with updated coordinates

Cluster 1 now comprises data points A, B, E, G. The updated centroid is **2.4**, **3.5**. Cluster 2 now comprises data points C, D, F. The updated centroid is **5**, **7**.

Step 6: Let's plug in our updated centroids on the scatterplot. You may notice that we are missing a data point. This is because the new centroid for Cluster 2 overlaps with data point D (5, 7), but this does not mean it has been removed or forgotten.

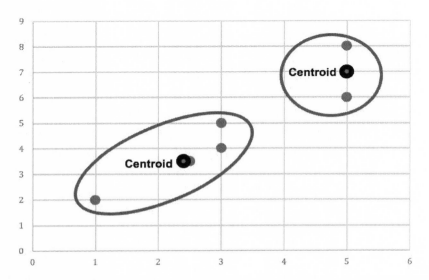

Figure 37: Two final clusters are produced based on the updated centroids for each cluster

In this iteration, each data point remains closest to its original cluster centroid and no data point has switched clusters. This provides us with our final result. Cluster 1 is A, B, E, G, and Cluster 2 is C, D, F.

For this example, it took two iterations to successfully create our two clusters. However, *k*-means clustering is not always able to reliably identify a final combination of clusters. In such cases, you will need to switch tactics and utilize another algorithm to formulate your classification model.

Regarding the limitations of cluster analysis, the most notable are multi-dimensional analysis and the identification of variable relevance. In this exercise, we looked at clustering performed on a two-dimensional scatterplot (based on two variables) but measuring multiple distances between data points in a three or four-dimensional space (with more than two variables) is much more complicated and time-consuming to compute.

The other downside of cluster analysis is that its success depends largely on the quality of data and there's no mechanism to differentiate between relevant and irrelevant variables. Clustering analysis segments data points based on the variables provided, but this isn't to say that the variables you selected are relevant and especially if chosen from a large pool of variables.

Let's consider the following scenario where we want to evaluate house properties in a new city with data based on these four variables:

1) Land size

2) House size

3) Distance from the CBD (central business district)

4) House price

To perform two-dimensional clustering analysis, we need to select two variables. However, there are no two true right variables. Sure, land size and house size have some degree of correlation and overlap, and perhaps one can be omitted. But, do we want to risk limiting our focus to houses that may have a backyard too large to manage or no backyard at all?

The answer depends on our needs and priorities. This is why the choice of variables included for clustering analysis is so vital; although cluster analysis efficiently splits data points into clusters, it can't read or relay back to you which variables are most relevant to decision-making. Instead, they assist in mapping natural groupings based on the variables you select.

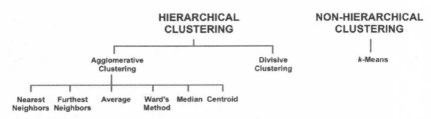

Figure 38: Summary of clustering types examined in this chapter

15

GATEWAY TO DATA SCIENCE

The modern development of computers and the transformative technology to collect, store and analyze data has made previously abstract methods of statistical analysis possible. However, rather than render statistics obsolete, modern technology and the emerging field of data science has energized research into statistical analysis as a way to analyze and solve the challenge of understanding big data.

While statistics and data science remain two distinct fields, there's a pronounced overlap on the data science side. This is because data science is underpinned by statistics and both fields draw heavily on collected data as input.

While an argument could be made that data science is a direct branch of statistics, that logic unravels when you consider the fact that machine learning, data mining, and big data analysis are not examined in pure statistics and these branches require skills outside of statistics. Thus, despite their relevance, statistics and data science remain distinct from one another and yet complementary for students that wish to study both fields. As the following diagram depicts, exposure to statistics and other fields can help to develop rewarding new skillsets.

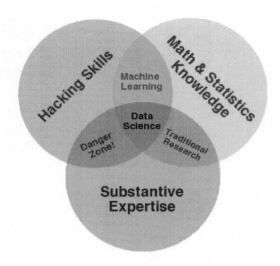

Figure 39: Conway's Venn Diagram, Source: http://drewconway.com/

This Venn diagram was created in 2010 by Drew Conway, a well-known data scientist and co-author of *Machine Learning for Hackers*. Using this visual diagram, Conway argues that data science requires a combination of three primary skills: hacking (coding), substantive expertise, and knowledge of math and statistics.

Math and statistics knowledge is a major component of Conway's diagram as statistical-based algorithms and math provide the guidelines for performing heavy number crunching. While coding can be applied to bypass the manual process of mathematical equations, a basic understanding of statistics is vital for working in data science. As shown in Conway's Venn diagram, hacking skills and domain expertise without sound knowledge of math and statistics presents a professional danger zone. Conway reasons that data scientists require at least a "baseline familiarity" of statistics and mathematics to select and interpret appropriate algorithms.[38]

---

[38] Drew Conway, "The Data Science Venn Diagram," *Drew Conway*, accessed December 12, 2017, http://drewconway.com/zia/2013/3/26/the-data-science-venn-diagram/.

There are other reasons why statistics is an important starting point for data scientists and especially as data science is more hands-on than we are led to believe. While modern technology has delivered drastic developments to mass-scale computation, data science is a multi-step process that relies on your ability to reason statistically. This includes statistical techniques such as random sampling. Advanced parallel computing is futile unless data is randomly selected and biases as obvious as the alphabetical listing of row values are rearranged.

The next step before any calculations are made involves managing the issue of missing values in the data. Something as simple and benign as a user skipping a field in an online form can cause headaches for data scientists who then have to approximate those missing values—especially for algorithms that don't accept missing values. Rather than delete all variables with a low number of missing values, data scientists can use central tendency measures (as introduced in Chapter 5) to approximate unknown values within the dataset.

The mean, median, and mode are three statistical techniques that can be used to plug gaps in the dataset. Each technique comes with its own bias of where the middle value lies, and while the mean and the median are effective for continuous variables such as money or physical weight, the mode is a better choice for variables with a low number of discrete or ordinal values, such as movie and product review ratings.

Statistical-based algorithms are then used prominently at the analysis stage. Although analytical packages and machine learning libraries such as Scikit-learn and TensorFlow integrate statistical algorithms into their routine code library, algorithmic parameters, random sampling, and missing values are not configured automatically. Through code, data scientists must manipulate the parameters of algorithms such as k-means clustering and regression analysis to optimize their model and reduce prediction error. Without a baseline familiarity of statistics, data scientists are vulnerable to misjudging algorithm settings and even misinterpreting probability—which is vital when looking for matches and associations. This also makes it easier to interpret and explain numbers produced by a computer algorithm to others, and as noted by Sarah Boslaugh in *Statistics in a Nutshell,* "it's always more fun to understand what you are doing."[39]

Looking to the Future

According to The British Broadcasting Company's (The BBC) interactive online resource *Will a robot take my job?*, statisticians (along with actuaries and economists) have a low threat of automation (15%) before the year 2035 and especially when compared with chartered accountants (95%), bank/post office clerks (97%), and receptionists (96%).[40]

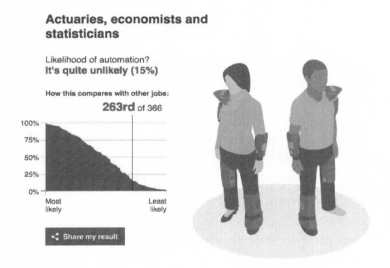

Figure 40: Likelihood of automation before the year 2035 for professional statisticians, Source: The BBC

---

[39] Sarah Boslaugh, "Statistics in a Nutshell," *O'Reilly Media*, Second Edition, 2012.
[40] "Will A Robot Take My Job?," *The BBC*, accessed December 30, 2016, http://www.bbc.com/news/technology-34066941/.

Thus, despite fears concerning job losses in other fields, statistics-focused job roles are well-positioned to ride the coming wave of AI and automation. Google's Chief Economist Hal Varian is one of the many figures buoyant about the prospects of the trained statistician.

"Data is so widely available and so strategically important that the scarce thing is the knowledge to extract wisdom from it. That is why statisticians, and database managers and machine learning people, are really going to be in a fantastic position."[41]

---

[41] Kenneth Cukier & Viktor Mayer-Schoenberger, "Big data: The critical ingredient," *Aljazeera*, accessed August 1, 2017, https://www.aljazeera.com/indepth/opinion/2013/11/big-data-critical-ingredient-201311182144196723.html/.

BUG BOUNTY

Thank you for reading this absolute beginner's introduction to statistics.

We offer a financial reward to readers for locating errors or bugs in this book. Some apparent errors could be mistakes made in interpreting a diagram or following along with the concepts introduced in the book, so we invite all readers to contact the author first for clarification and a possible reward, before posting a one-star review! Just send an email to **oliver.theobald@scatterplotpress.com** explaining the error or mistake you encountered.

This way, we can also supply further explanations and examples over email to calibrate your understanding, or in cases where you're right, and we're wrong, we offer a monetary reward through PayPal or Amazon gift card. This way you can make a tidy profit from your feedback, and we can update the book to improve the standard of content for future readers.

FURTHER RESOURCES

The Signal and the Noise: The Art and Science of Prediction

Format: E-book, Book

Author: Nate Silver

An excellent look at applied statistics (including Bayes' theorem) for a general readership who wants to improve their ability to make predictions; especially for anyone interested in science, economics, and sports betting.

Borel

Format: Board Game

A clever board game (co-designed by a statistician) in which players navigate chance and randomness with probabilistic reasoning to bet on the outcomes of simple experiments.

A Field Guide to Lies and Statistics: A Neuroscientist on How to Make Sense of a Complex World

Format: Book

Author: Daniel Levitin

Despite the mention of "Neuroscientist" in the title, the exploration of methods in this book is, in fact, ideal for beginners. Reading this book will help you to avoid some of the common pitfalls of statistical analysis and likely prompt you to scrutinize facts and figures regularly reported in the media.

Hypothesis Testing: A Visual Introduction To Statistical Significance

Format: E-book

Author: Scott Hartshorn

A short, affordable (USD $3.20), and engaging read on hypothesis testing with detailed visual examples, useful practical tips, and clear instructions.

Machine Learning for Absolute Beginners, Second Edition

Format: E-book, Book

Author: Oliver Theobald

As the first published work of this book's author, *Machine Learning for Absolute Beginners* provides a clear and high-level introduction to machine learning including numerous algorithms and practical examples to help you build your first machine learning model.

Linear Regression And Correlation: A Beginner's Guide

Format: E-book

Author: Scott Hartshorn

Suggested Audience: All

A well-explained and affordable (USD $3.20) introduction to linear regression and correlation.

The Basement Tapes, Revisionist History, Series 2, Episode 10

Format: Podcast

Host: Malcolm Gladwell

In this podcast, best-selling author Malcolm Gladwell explores the story of a cardiologist in Minnesota who finds a box of data from an experiment in his basement that will later change our understanding of the modern American diet. This story places hypothesis testing and other concepts introduced in this book into a real-life context.

Intro to Statistics

Format: Udacity course

As a free beginner's course to statistics on Udacity, this online resource introduces techniques for visualizing relationships in data and systematic techniques for understanding relationships using mathematics.

Naked Statistics: Stripping the Dread from the Data

Format: E-book, Book

Author: Charles Wheelan

Easy to read and designed to be entertaining, this book clarifies key concepts including inference, correlation, and regression analysis to solve practical problems. This book, though, does skip over a lot of the specifics, which may not be useful for those past the absolute beginner stage.

Famous Statistical Quotations

Format: Stack Exchange Thread

Author: Multiple sources from Stack Exchange community

A list of famous statistical quotations up-voted by contributors on Stack Exchange that you might like to jot down on post-it notes and decorate your cubicle. Popular quotations on the site include:

"I think it's much more interesting to live not knowing than to have answers which might be wrong" —Richard Feynman

"Statistics are like bikinis. What they reveal is suggestive, but what they conceal is vital" —Aaron Levenstein

"Statistical thinking will one day be as necessary a qualification for efficient citizenship as the ability to read and write" —H.G. Wells

OTHER BOOKS BY THE AUTHOR

Made in the USA
Middletown, DE
04 March 2021

34724825R00094